SECOND EDITION

Food Plant
SANITATION

Design, Maintenance, and Good Manufacturing Practices

MICHAEL M. CRAMER

CRC Press
Taylor & Francis Group
Boca Raton   London   New York

CRC Press is an imprint of the
Taylor & Francis Group, an **informa** business

CRC Press
Taylor & Francis Group
6000 Broken Sound Parkway NW, Suite 300
Boca Raton, FL 33487-2742

First issued in paperback 2016

© 2013 by Taylor & Francis Group, LLC
CRC Press is an imprint of Taylor & Francis Group, an Informa business

No claim to original U.S. Government works

Version Date: 20130404

ISBN 13: 978-1-138-19879-1 (pbk)
ISBN 13: 978-1-4665-1173-6 (hbk)

Library of Congress Cataloging-in-Publication Data

Cramer, Michael M.
    Food plant sanitation : design, maintenance, and good manufacturing practices / author, Michael M. Cramer. -- Second edition.
        pages cm
        Includes bibliographical references and index.
        ISBN 978-1-4665-1173-6 (hardback)
        1. Food industry and trade--Sanitation. I. Title.

TP373.6.C73 2013
664'.07--dc23   2013010830

Visit the Taylor & Francis Web site at
http://www.taylorandfrancis.com

and the CRC Press Web site at
http://www.crcpress.com

Contents

Preface

Food manufacturing companies have both a legal and an ethical responsibility to provide the consuming public with foods that are safe and wholesome. Production of foods that are safe for consumption and meet consumer expectations for quality and palatability also makes a good business sense as it will encourage consumers to patronize those products repeatedly over time. As food safety and quality professionals, we have a responsibility to provide leadership to our respective companies in the area of food sanitation and demonstrate how it relates to meeting all food safety, quality, and productivity goals. We also have a duty to share our knowledge of sanitation and safe practices with others in the industry. It is for this reason that this book has been written, so that others may benefit from improved business and increased consumer satisfaction and confidence.

This book is also intended to provide practical advice on all aspects of food plant sanitation and sanitation-related food safety issues. It is intended to provide the reader with the tools to establish a food safety system to aid in control of microbiological, physical, and chemical hazards. If a company begins with the basic understanding that sanitation is integral to food safety, then they have set the foundation for an effective food safety system. As such, this book provides some of the key components to that system: a description of the recent challenges faced by the industry due to pathogens such as *Listeria monocytogenes*, biofilms and allergens, proven industry best practices for sanitation in clear and simple terms as well as current sanitary regulatory requirements from both the FDA and USDA.

I have thoroughly enjoyed working in the food industry for more than 30 years and have learned from some of the leaders and experts from the industry, academia, and regulatory agencies. I have also learned from my mistakes and have attempted to provide a practical perspective on implementation of proven food plant sanitation and safety processes. Where possible, I have also included examples of procedures, forms, and documents that can aid novice food safety and quality professionals with the development of their food safety systems. If the use of the material in this book prevents one food-related illness or injury, allows a company to avoid regulatory control action or recall, or provides the user with the tools to improve product performance, then writing this book has been more than a labor of love; it has been worthwhile.

The Author

Michael Cramer is a native of New York state. He started his food career at a turkey processing facility in eastern Pennsylvania while attending West Chester University in West Chester, Pennsylvannia. Following graduation in May 1977 with a Bachelor of Science degree in Health Education, Cramer went back to work in the turkey plant through the remainder of the processing season. He was hired by Swift & Company as Quality Assurance Manager at the facility in April 1978. In 1979 he was transferred to a larger facility in North Carolina, where he gained experience in turkey breeding, feed mill and hatchery operations, and further processing. He transferred to Swift & Company corporate headquarters in Illinois in 1984, where he held positions as a Production Specialist and Documentation Manager. In the position as Documentation Manager, he helped develop ingredient specifications and product manufacturing documents and managed the regulatory process for USDA label submission. He transferred to the Swift-Eckrich plant in downtown Chicago as plant QA Manager in 1989, where he spent two years working with cooked sausages and hotdogs. He returned to the corporate office in Downers Grove, Illinois, in 1991 as QA Director for the Armour food plants of the newly formed Armour Swift-Eckrich. Here he implemented food safety and quality programs for cooked ready-to-eat meat and poultry products, dry and semi-dry sausage, and cooked meat and poultry patties.

In 1992 Cramer accepted a position as Quality Assurance Director for Newly Weds Foods in Chicago, Illinois, working with batter and breading coating systems and spice blending operations. Cramer left Newly Weds in 1993 for the position of Quality Assurance Director for Specialty Brands Inc. in Southern California. Here he developed and implemented a comprehensive food safety and quality system that included HACCP, SSOP, Allergen Management, Environmental Microbiological Management, and a quality system that focused on identification of quality control points in the production process. He was promoted to Vice President Food Safety and Quality Assurance in 1999.

In 2004 Cramer formed Cramer & Associates Food Safety Services, consulting with food manufacturing plants across the United States and partnering closely with Food Safety Net Services and the HACCP Consulting Group. Cramer joined

Windsor Foods, one of his largest clients, as Quality Assurance Director in 2005 with responsibility for their western operations. In 2012 he was promoted to Sr. Director, Food Safety & Quality Assurance, with responsibility for all 11 Windsor Foods operations.

Cramer is a Professional Member of the Institute of Food Technologists, a member of the International Association for Food Protection, on the editorial board of *Food Safety Magazine*, and a former Director with the National Meat Association. He has written articles for *Food Safety Magazine* dealing with Biosecurity, Sanitation and Sanitary Design, and Allergens. He has been a regular presenter at the annual National Meat Association and Southwest Meat Association conferences as well as at the annual Food Safety Summit in Washington, DC.

He lists his influences starting with his parents, who have shown him tremendous love and support; his sisters, who have been his role models and sources of inspiration for their great career and family achievements; and his nephews, who remind him that you are never too old to play and have fun! There have been many great influences in his life and career, including teachers such as Ed Vozella, whom he credits with influencing him to write; industry leaders such as Bruce Tompkin, who has forgotten more about microbiology and food safety than Cramer will ever know; and Rosemary Mucklow, who has been an amazing influence in the food industry. He would also like to credit the many friends who have provided support and encouragement and who have taught him more than they ever know.

Cramer lives in Southern California where he enjoys skiing, hiking, camping, biking, fine wine, food, and travel. He looks forward to returning to Ireland one day.

Chapter 1

Sanitation Regulatory Requirements

> ... all of the old smells of a generation would be drawn out by this heat—for there was never any washing of the walls and rafters and pillars, and they were caked with the filth of a lifetime.
>
> *The Jungle* **by Upton Sinclair**

Upton Sinclair's book *The Jungle* was considered by many to be a general reflection of unsanitary and unsafe working conditions in the meatpacking houses of the early 20th century. It is believed that this story ultimately led President Theodore Roosevelt to push for many of the federal regulations, including the Meat Inspection Act of 1906 and the Pure Food Act, that apply to the food manufacturing industry today. The intent of food laws is to protect the public because consumers cannot detect contamination in food simply by sight, smell, taste, or touch. Consumers must rely on food manufacturers and other parties, including the government, to ensure that they are provided with safe food products [6]; consequently, Congress assumed responsibility for food protection, and though the industry has changed significantly for the better, the regulations that apply to food manufacturing are done so with the intention of protecting the consuming public and plant workers. Much of *The Jungle* centers around the sanitary working conditions in food manufacturing, and manufacturers now recognize that sanitary operations are not only required by law but make for good business practices. As a result of the implementation of these laws and through the efforts by the food industry, the United States is recognized as having the safest food supply in the world.

Food plants may operate under federal regulations or various state and local codes. All are designed to prevent production of food ingredients or products that may lead to contamination with filth, hazardous substances, or adulteration. Whether operating under Food and Drug Administration (FDA) or United States Department of Agriculture (USDA), food plants are governed by federal regulations. The intent of this chapter is to present in simple terms the basics of federal regulations covering sanitation and sanitary operations. In addition to FDA and USDA, other agencies concerned with sanitation are the Environmental Protection Agency (EPA) (FIFRA, RCRA, Clean Air Act, Federal Water Pollution Control Act) and OSHA for worker safety, and state and local public health agencies [5].

FDA

In 1938, Congress passed the Pure Food, Drug and Cosmetic Act that covers all commodities, except meat and poultry products, for which USDA has responsibility, in interstate commerce from harvest through processing and distribution. It is the principal statute covering production and distribution of foods in the United States [6]. The act establishes tolerances for unavoidable toxic substances, authorizes factory inspections, and declares that food is adulterated if it has been prepared, packed, or stored under conditions in which it might have been contaminated. In 1969, the first Good Manufacturing Practices (GMPs) were published as Part 128 of the Code of Federal Regulations (CFR) [4]. The CFR is divided into approximately 50 titles, and Title 21 deals with food and drugs. The GMPs were recodified in 1977 as Part 110 of the CFR and revised and updated in 1986, at which time a reference to HACCP was included and stipulated a need to control "undesirable microorganisms" [7]. They explain GMP requirements that cover all aspects of food manufacturing from employee requirements, through facility and equipment design and cleaning. Implementation of GMPs in a food operation will be covered in greater detail in Chapter 7 of this book. The sections of Part 110 are reprinted below [1].

Title 21—Food and Drugs

Chapter I—Food and Drug Administration, Department of Health and Human Services (continued)

Part 110_Current Good Manufacturing Practice in Manufacturing, Packing, or Holding Human Food—Table of Contents

Subpart A General Provisions

Part 110.5 Current Good Manufacturing Practices
  (a) The criteria and definitions in this part shall apply in determining whether a food is adulterated (1) within the meaning of

section 402(a) (3) of the act in that the food has been manufactured under such conditions that it is unfit for food; or (2)
within the meaning of section 402(a)(4) of the act in that the
food has been prepared, packed, or held under insanitary conditions whereby it may have become contaminated with filth, or
whereby it may have been rendered injurious to health. The criteria and definitions in this part also apply in determining whether
a food is in violation of section 361 of the Public Health Service
Act (42 U.S.C. 264).

(b) Food covered by specific current good manufacturing practice
regulations also is subject to the requirements of those regulations.

Part 110.10 Personnel   The plant management shall take all reasonable
measures and precautions to ensure the following:

(a) Disease control. Any person who, by medical examination or
supervisory observation, is shown to have, or appears to have, an
illness, open lesion, including boils, sores, or infected wounds, or
any other abnormal source of microbial contamination by which
there is a reasonable possibility of food, food-contact surfaces,
or food-packaging materials becoming contaminated, shall be
excluded from any operations which may be expected to result
in such contamination until the condition is corrected. Personnel
shall be instructed to report such health conditions to their
supervisors.

(b) Cleanliness. All persons working in direct contact with food,
food-contact surfaces, and food-packaging materials shall conform to hygienic practices while on duty to the extent necessary
to protect against contamination of food. The methods for maintaining cleanliness include, but are not limited to:
   (1) Wearing outer garments suitable to the operation in a manner
that protects against the contamination of food, food-contact
surfaces, or food-packaging materials.
   (2) Maintaining adequate personal cleanliness.
   (3) Washing hands thoroughly (and sanitizing if necessary to
protect against contamination with undesirable microorganisms) in an adequate hand-washing facility before starting work, after each absence from the work station, and at
any other time when the hands may have become soiled or
contaminated.
   (4) Removing all unsecured jewelry and other objects that might
fall into food, equipment, or containers, and removing hand

jewelry that cannot be adequately sanitized during periods in which food is manipulated by hand. If such hand jewelry cannot be removed, it may be covered by material which can be maintained in an intact, clean, and sanitary condition and which effectively protects against the contamination by these objects of the food, food-contact surfaces, or food-packaging materials.

(5) Maintaining gloves, if they are used in food handling, in an intact, clean, and sanitary condition. The gloves should be of an impermeable material.

(6) Wearing, where appropriate, in an effective manner, hair nets, headbands, caps, beard covers, or other effective hair restraints.

(7) Storing clothing or other personal belongings in areas other than where food is exposed or where equipment or utensils are washed.

(8) Confining the following to areas other than where food may be exposed or where equipment or utensils are washed: eating food, chewing gum, drinking beverages, or using tobacco.

(9) Taking any other necessary precautions to protect against contamination of food, food-contact surfaces, or food-packaging materials with microorganisms or foreign substances including, but not limited to, perspiration, hair, cosmetics, tobacco, chemicals, and medicines applied to the skin.

(c) Education and training. Personnel responsible for identifying sanitation failures or food contamination should have a background of education or experience, or a combination thereof, to provide a level of competency necessary for production of clean and safe food. Food handlers and supervisors should receive appropriate training in proper food handling techniques and food-protection principles and should be informed of the danger of poor personal hygiene and insanitary practices.

(d) Supervision. Responsibility for assuring compliance by all personnel with all requirements of this part shall be clearly assigned to competent supervisory personnel.

Part 110.20 Plant and Grounds

(a) Grounds. The grounds about a food plant under the control of the operator shall be kept in a condition that will protect against the contamination of food. The methods for adequate maintenance of grounds include, but are not limited to:

(1) Properly storing equipment, removing litter and waste, and cutting weeds or grass within the immediate vicinity of the

plant buildings or structures that may constitute an attractant, breeding place, or harborage for pests.

(2) Maintaining roads, yards, and parking lots so that they do not constitute a source of contamination in areas where food is exposed.

(3) Adequately draining areas that may contribute contamination to food by seepage, foot-borne filth, or providing a breeding place for pests.

(4) Operating systems for waste treatment and disposal in an adequate manner so that they do not constitute a source of contamination in areas where food is exposed.

If the plant grounds are bordered by grounds not under the operator's control and not maintained in the manner described in paragraph (a) (1) through (3) of this section, care shall be exercised in the plant by inspection, extermination, or other means to exclude pests, dirt, and filth that may be a source of food contamination.

(b) Plant construction and design. Plant buildings and structures shall be suitable in size, construction, and design to facilitate maintenance and sanitary operations for food-manufacturing purposes. The plant and facilities shall:

(1) Provide sufficient space for such placement of equipment and storage of materials as is necessary for the maintenance of sanitary operations and the production of safe food.

(2) Permit the taking of proper precautions to reduce the potential for contamination of food, food-contact surfaces, or food-packaging materials with microorganisms, chemicals, filth, or other extraneous material. The potential for contamination may be reduced by adequate food safety controls and operating practices or effective design, including the separation of operations in which contamination is likely to occur, by one or more of the following means: location, time, partition, airflow, enclosed systems, or other effective means.

(3) Permit the taking of proper precautions to protect food in outdoor bulk fermentation vessels by any effective means, including:

(i) Using protective coverings.

(ii) Controlling areas over and around the vessels to eliminate harborages for pests.

(iii) Checking on a regular basis for pests and pest infestation.

(iv) Skimming the fermentation vessels, as necessary.

(4) Be constructed in such a manner that floors, walls, and ceilings may be adequately cleaned and kept clean and kept in

good repair; that drip or condensate from fixtures, ducts and pipes does not contaminate food, food-contact surfaces, or food-packaging materials; and that aisles or working spaces are provided between equipment and walls and are adequately unobstructed and of adequate width to permit employees to perform their duties and to protect against contaminating food or food-contact surfaces with clothing or personal contact.

(5) Provide adequate lighting in hand-washing areas, dressing and locker rooms, and toilet rooms and in all areas where food is examined, processed, or stored and where equipment or utensils are cleaned; and provide safety-type light bulbs, fixtures, skylights, or other glass suspended over exposed food in any step of preparation or otherwise protect against food contamination in case of glass breakage.

(6) Provide adequate ventilation or control equipment to minimize odors and vapors (including steam and noxious fumes) in areas where they may contaminate food; and locate and operate fans and other air-blowing equipment in a manner that minimizes the potential for contaminating food, food-packaging materials, and food-contact surfaces.

(7) Provide, where necessary, adequate screening or other protection against pests.

Part 110.35 Sanitary Operations

(a) General maintenance. Buildings, fixtures, and other physical facilities of the plant shall be maintained in a sanitary condition and shall be kept in repair sufficient to prevent food from becoming adulterated within the meaning of the act. Cleaning and sanitizing of utensils and equipment shall be conducted in a manner that protects against contamination of food, food-contact surfaces, or food-packaging materials.

(b) Substances used in cleaning and sanitizing; storage of toxic materials.

(1) Cleaning compounds and sanitizing agents used in cleaning and sanitizing procedures shall be free from undesirable microorganisms and shall be safe and adequate under the conditions of use. Compliance with this requirement may be verified by any effective means including purchase of these substances under a supplier's guarantee or certification, or examination of these substances for contamination. Only the following toxic materials may be used or stored in a plant where food is processed or exposed:

(i) Those required to maintain clean and sanitary conditions;

(ii) Those necessary for use in laboratory testing procedures;

(iii) Those necessary for plant and equipment maintenance and operation; and

(iv) Those necessary for use in the plant's operations.

(2) Toxic cleaning compounds, sanitizing agents, and pesticide chemicals shall be identified, held, and stored in a manner that protects against contamination of food, food-contact surfaces, or food-packaging materials. All relevant regulations promulgated by other Federal, State, and local government agencies for the application, use, or holding of these products should be followed.

(c) Pest control. No pests shall be allowed in any area of a food plant. Guard or guide dogs may be allowed in some areas of a plant if the presence of the dogs is unlikely to result in contamination of food, food contact surfaces, or food-packaging materials. Effective measures shall be taken to exclude pests from the processing areas and to protect against the contamination of food on the premises by pests. The use of insecticides or rodenticides is permitted only under precautions and restrictions that will protect against the contamination of food, food-contact surfaces, and food-packaging materials.

(d) Sanitation of food-contact surfaces. All food-contact surfaces, including utensils and food-contact surfaces of equipment, shall be cleaned as frequently as necessary to protect against contamination of food.

(1) Food-contact surfaces used for manufacturing or holding low-moisture food shall be in a dry, sanitary condition at the time of use.

When the surfaces are wet-cleaned, they shall, when necessary, be sanitized and thoroughly dried before subsequent use.

(2) In wet processing, when cleaning is necessary to protect against the introduction of microorganisms into food, all food-contact surfaces shall be cleaned and sanitized before use and after any interruption during which the food-contact surfaces may have become contaminated. Where equipment and utensils are used in a continuous production operation, the utensils and food-contact surfaces of the equipment shall be cleaned and sanitized as necessary.

(3) Non-food-contact surfaces of equipment used in the operation of food plants should be cleaned as frequently as necessary to protect against contamination of food.

(4) Single-service articles (such as utensils intended for one-time use, paper cups, and paper towels) should be stored in appropriate containers and shall be handled, dispensed, used, and

disposed of in a manner that protects against contamination of food or food-contact surfaces.

(5) Sanitizing agents shall be adequate and safe under conditions of use. Any facility, procedure, or machine is acceptable for cleaning and sanitizing equipment and utensils if it is established that the facility, procedure, or machine will routinely render equipment and utensils clean and provide adequate cleaning and sanitizing treatment.

(e) Storage and handling of cleaned portable equipment and utensils. Cleaned and sanitized portable equipment with food-contact surfaces and utensils should be stored in a location and manner that protects food-contact surfaces from contamination.

Part 110.37 Sanitary Facilities and Controls   Each plant shall be equipped with adequate sanitary facilities and accommodations including, but not limited to:

(a) Water supply. The water supply shall be sufficient for the operations intended and shall be derived from an adequate source. Any water that contacts food or food-contact surfaces shall be safe and of adequate sanitary quality. Running water at a suitable temperature, and under pressure as needed, shall be provided in all areas where required for the processing of food, for the cleaning of equipment, utensils, and food-packaging materials, or for employee sanitary facilities.

(b) Plumbing. Plumbing shall be of adequate size and design and adequately installed and maintained to:
 (1) Carry sufficient quantities of water to required locations throughout the plant.
 (2) Properly convey sewage and liquid disposable waste from the plant.
 (3) Avoid constituting a source of contamination to food, water supplies, equipment, or utensils or creating an unsanitary condition.
 (4) Provide adequate floor drainage in all areas where floors are subject to flooding-type cleaning or where normal operations release or discharge water or other liquid waste on the floor.
 (5) Provide that there is not backflow from, or cross-connection between, piping systems that discharge waste water or sewage and piping systems that carry water for food or food manufacturing.

(c) Sewage disposal. Sewage disposal shall be made into an adequate sewerage system or disposed of through other adequate means.

(d) Toilet facilities. Each plant shall provide its employees with adequate, readily accessible toilet facilities. Compliance with this requirement may be accomplished by:
  (1) Maintaining the facilities in a sanitary condition.
  (2) Keeping the facilities in good repair at all times.
  (3) Providing self-closing doors.
  (4) Providing doors that do not open into areas where food is exposed to airborne contamination, except where alternate means have been taken to protect against such contamination (such as double doors or positive airflow systems).
(e) Hand-washing facilities. Hand-washing facilities shall be adequate and convenient and be furnished with running water at a suitable temperature. Compliance with this requirement may be accomplished by providing:
  (1) Hand-washing and, where appropriate, hand-sanitizing facilities at each location in the plant where good sanitary practices require employees to wash and/or sanitize their hands.
  (2) Effective hand-cleaning and sanitizing preparations.
  (3) Sanitary towel service or suitable drying devices.
  (4) Devices or fixtures, such as water control valves, so designed and constructed to protect against recontamination of clean, sanitized hands.
  (5) Readily understandable signs directing employees handling unprotected food, unprotected food-packaging materials, of food-contact surfaces to wash and, where appropriate, sanitize their hands before they start work, after each absence from post of duty, and when their hands may have become soiled or contaminated. These signs may be posted in the processing room(s) and in all other areas where employees may handle such food, materials, or surfaces.
  (6) Refuse receptacles that are constructed and maintained in a manner that protects against contamination of food.
(f) Rubbish and offal disposal. Rubbish and any offal shall be so conveyed, stored, and disposed of as to minimize the development of odor, minimize the potential for the waste becoming an attractant and harborage or breeding place for pests, and protect against contamination of food, food-contact surfaces, water supplies, and ground surfaces.

Part 110.40 Equipment and Utensils
  (a) All plant equipment and utensils shall be so designed and of such material and workmanship as to be adequately cleanable, and

shall be properly maintained. The design, construction, and use of equipment and utensils shall preclude the adulteration of food with lubricants, fuel, metal fragments, contaminated water, or any other contaminants. All equipment should be so installed and maintained as to facilitate the cleaning of the equipment and of all adjacent spaces. Food-contact surfaces shall be corrosion-resistant when in contact with food. They shall be made of nontoxic materials and designed to withstand the environment of their intended use and the action of food, and, if applicable, cleaning compounds and sanitizing agents. Food-contact surfaces shall be maintained to protect food from being contaminated by any source, including unlawful indirect food additives.

(b) Seams on food-contact surfaces shall be smoothly bonded or maintained so as to minimize accumulation of food particles, dirt, and organic matter and thus minimize the opportunity for growth of microorganisms.

(c) Equipment that is in the manufacturing or food-handling area and that does not come into contact with food shall be so constructed that it can be kept in a clean condition.

(d) Holding, conveying, and manufacturing systems, including gravimetric, pneumatic, closed, and automated systems, shall be of a design and construction that enables them to be maintained in an appropriate sanitary condition.

(e) Each freezer and cold storage compartment used to store and hold food capable of supporting growth of microorganisms shall be fitted with an indicating thermometer, temperature-measuring device, or temperature-recording device so installed as to show the temperature accurately within the compartment, and should be fitted with an automatic control for regulating temperature or with an automatic alarm system to indicate a significant temperature change in a manual operation.

(f) Instruments and controls used for measuring, regulating, or recording temperatures, pH, acidity, water activity, or other conditions that control or prevent the growth of undesirable microorganisms in food shall be accurate and adequately maintained, and adequate in number for their designated uses.

(g) Compressed air or other gases mechanically introduced into food or used to clean food-contact surfaces or equipment shall be treated in such a way that food is not contaminated with unlawful indirect food additives.

Part 110.80 Process and Controls   All operations in the receiving, inspecting, transporting, segregating, preparing, manufacturing, packaging, and storing of food shall be conducted in accordance with adequate sanitation principles. Appropriate quality control operations shall be employed to ensure that food is suitable for human consumption and that food-packaging materials are safe and suitable. Overall sanitation of the plant shall be under the supervision of one or more competent individuals assigned responsibility for this function. All reasonable precautions shall be taken to ensure that production procedures do not contribute contamination from any source. Chemical, microbial, or extraneous-material testing procedures shall be used where necessary to identify sanitation failures or possible food contamination. All food that has become contaminated to the extent that it is adulterated within the meaning of the act shall be rejected, or if permissible, treated or processed to eliminate the contamination.

(a) Raw materials and other ingredients.
   (1) Raw materials and other ingredients shall be inspected and segregated or otherwise handled as necessary to ascertain that they are clean and suitable for processing into food and shall be stored under conditions that will protect against contamination and minimize deterioration. Raw materials shall be washed or cleaned as necessary to remove soil or other contamination. Water used for washing, rinsing, or conveying food shall be safe and of adequate sanitary quality. Water may be reused for washing, rinsing, or conveying food if it does not increase the level of contamination of the food. Containers and carriers of raw materials should be inspected on receipt to ensure that their condition has not contributed to the contamination or deterioration of food.
   (2) Raw materials and other ingredients shall either not contain levels of microorganisms that may produce food poisoning or other disease in humans, or they shall be pasteurized or otherwise treated during manufacturing operations so that they no longer contain levels that would cause the product to be adulterated within the meaning of the act. Compliance with this requirement may be verified by any effective means, including purchasing raw materials and other ingredients under a supplier's guarantee or certification.
   (3) Raw materials and other ingredients susceptible to contamination with aflatoxin or other natural toxins shall comply with current Food and Drug Administration regulations and action levels for poisonous or deleterious substances before these materials or ingredients are incorporated into finished food. Compliance with this requirement may be

accomplished by purchasing raw materials and other ingredients under a supplier's guarantee or certification, or may be verified by analyzing these materials and ingredients for aflatoxins and other natural toxins.

(4) Raw materials, other ingredients, and rework susceptible to contamination with pests, undesirable microorganisms, or extraneous material shall comply with applicable Food and Drug Administration regulations and defect action levels for natural or unavoidable defects if a manufacturer wishes to use the materials in manufacturing food.

Compliance with this requirement may be verified by any effective means, including purchasing the materials under a supplier's guarantee or certification, or examination of these materials for contamination.

(5) Raw materials, other ingredients, and rework shall be held in bulk, or in containers designed and constructed so as to protect against contamination and shall be held at such temperature and relative humidity and in such a manner as to prevent the food from becoming adulterated within the meaning of the act. Material scheduled for rework shall be identified as such.

(6) Frozen raw materials and other ingredients shall be kept frozen. If thawing is required prior to use, it shall be done in a manner that prevents the raw materials and other ingredients from becoming adulterated within the meaning of the act.

(7) Liquid or dry raw materials and other ingredients received and stored in bulk form shall be held in a manner that protects against contamination.

(b) Manufacturing operations.

(1) Equipment and utensils and finished food containers shall be maintained in an acceptable condition through appropriate cleaning and sanitizing, as necessary. Insofar as necessary, equipment shall be taken apart for thorough cleaning.

(2) All food manufacturing, including packaging and storage, shall be conducted under such conditions and controls as are necessary to minimize the potential for the growth of microorganisms, or for the contamination of food. One way to comply with this requirement is careful monitoring of physical factors such as time, temperature, humidity, aw, pH, pressure, flow rate, and manufacturing operations such as freezing, dehydration, heat processing, acidification, and

refrigeration to ensure that mechanical breakdowns, time delays, temperature fluctuations, and other factors do not contribute to the decomposition or contamination of food.

(3) Food that can support the rapid growth of undesirable microorganisms, particularly those of public health significance, shall be held in a manner that prevents the food from becoming adulterated within the meaning of the act. Compliance with this requirement may be accomplished by any effective means, including:

    (i)   Maintaining refrigerated foods at 45 [deg]F (7.2 [deg]C) or below as appropriate for the particular food involved.

    (ii) Maintaining frozen foods in a frozen state.

    (iii) Maintaining hot foods at 140 [deg]F (60 [deg]C) or above.

    (iv) Heat treating acid or acidified foods to destroy mesophilic microorganisms when those foods are to be held in hermetically sealed containers at ambient temperatures.

(4) Measures such as sterilizing, irradiating, pasteurizing, freezing, refrigerating, controlling pH or controlling aw that are taken to destroy or prevent the growth of undesirable microorganisms, particularly those of public health significance, shall be adequate under the conditions of manufacture, handling, and distribution to prevent food from being adulterated within the meaning of the act.

(5) Work-in-process shall be handled in a manner that protects against contamination.

(6) Effective measures shall be taken to protect finished food from contamination by raw materials, other ingredients, or refuse. When raw materials, other ingredients, or refuse are unprotected, they shall not be handled simultaneously in a receiving, loading, or shipping area if that handling could result in contaminated food. Food transported by conveyor shall be protected against contamination as necessary.

(7) Equipment, containers, and utensils used to convey, hold, or store raw materials, work-in-process, rework, or food shall be constructed, handled, and maintained during manufacturing or storage in a manner that protects against contamination.

(8) Effective measures shall be taken to protect against the inclusion of metal or other extraneous material in food. Compliance with this requirement may be accomplished by using sieves, traps, magnets, electronic metal detectors, or other suitable effective means.

(9) Food, raw materials, and other ingredients that are adulterated within the meaning of the act shall be disposed of in a manner that protects against the contamination of other food. If the adulterated food is capable of being reconditioned, it shall be reconditioned using a method that has been proven to be effective or it shall be reexamined and found not to be adulterated within the meaning of the act before being incorporated into other food.

(10) Mechanical manufacturing steps such as washing, peeling, trimming, cutting, sorting and inspecting, mashing, dewatering, cooling, shredding, extruding, drying, whipping, defatting, and forming shall be performed so as to protect food against contamination. Compliance with this requirement may be accomplished by providing adequate physical protection of food from contaminants that may drip, drain, or be drawn into the food. Protection may be provided by adequate cleaning and sanitizing of all food-contact surfaces, and by using time and temperature controls at and between each manufacturing step.

(11) Heat blanching, when required in the preparation of food, should be effected by heating the food to the required temperature, holding it at this temperature for the required time, and then either rapidly cooling the food or passing it to subsequent manufacturing without delay. Thermophilic growth and contamination in blanchers should be minimized by the use of adequate operating temperatures and by periodic cleaning. Where the blanched food is washed prior to filling, water used shall be safe and of adequate sanitary quality.

(12) Batters, breading, sauces, gravies, dressings, and other similar preparations shall be treated or maintained in such a manner that they are protected against contamination. Compliance with this requirement may be accomplished by any effective means, including one or more of the following:
   (i) Using ingredients free of contamination.
   (ii) Employing adequate heat processes where applicable.
   (iii) Using adequate time and temperature controls.
   (iv) Providing adequate physical protection of components from contaminants that may drip, drain, or be drawn into them.
   (v) Cooling to adequate temperature during manufacturing.
   (vi) Disposing of batters at appropriate intervals to protect against the growth of microorganisms.

(13) Filling, assembling, packaging, and other operations shall be performed in such a way that the food is protected against contamination. Compliance with this requirement may be accomplished by any effective means, including:

   (i) Use of a quality control operation in which the critical control points are identified and controlled during manufacturing.

   (ii) Adequate cleaning and sanitizing of all food-contact surfaces and food containers.

   (iii) Using materials for food containers and food-packaging materials that are safe and suitable, as defined in Sec. 130.3(d) of this chapter.

   (iv) Providing physical protection from contamination, particularly airborne contamination.

   (v) Using sanitary handling procedures.

(14) Food such as, but not limited to, dry mixes, nuts, intermediate moisture food, and dehydrated food, that relies on the control of aw for preventing the growth of undesirable microorganisms shall be processed to and maintained at a safe moisture level. Compliance with this requirement may be accomplished by any effective means, including employment of one or more of the following practices:

   (i) Monitoring the aw of food.

   (ii) Controlling the soluble solids-water ratio in finished food.

   (iii) Protecting finished food from moisture pickup, by use of a moisture barrier or by other means, so that the aw of the food does not increase to an unsafe level.

(15) Food such as, but not limited to, acid and acidified food, that relies principally on the control of pH for preventing the growth of undesirable microorganisms shall be monitored and maintained at a pH of 4.6 or below. Compliance with this requirement may be accomplished by any effective means, including employment of one or more of the following practices:

   (i) Monitoring the pH of raw materials, food in process, and finished food.

   (ii) Controlling the amount of acid or acidified food added to low-acid food.

(16) When ice is used in contact with food, it shall be made from water that is safe and of adequate sanitary quality, and shall be used only if it has been manufactured in accordance with current good manufacturing practice as outlined in this part.

(17) Food-manufacturing areas and equipment used for manufacturing human food should not be used to manufacture

nonhuman food-grade animal feed or inedible products, unless there is no reasonable possibility for the contamination of the human food.

Part 110.93 Warehousing and Distribution   Storage and transportation of finished food shall be under conditions that will protect food against physical, chemical, and microbial contamination as well as against deterioration of the food and the container.

These sections as identified should be used by food manufacturers as a template and guide to designing their general sanitation and food safety systems to ensure compliance, but more important to prevent the production of adulterated food and to protect the consumer.

Food Safety Modernization Act of 2011

The act was signed into law in 2011 and marks the first major overhaul of food safety legislation since the passage of the Pure Food, Drug and Cosmetic Act. Passage of the Food Safety Modernization Act (FSMA) was believed to be, in large part, due to the massive peanut butter recall of 2009. {10} The act provides the agency with expanded powers and is intended to move FDA toward prevention-oriented food safety systems.

The act is broken down into 41 sections. While none of the sections refers specifically to food manufacturing plant sanitation, they will be of interest to both domestic and international food manufacturers and those who import and export foods. Of significant importance are the following sections: Registration of food facilities (Section 102) under FDA inspection in 2012 and subsequent re-registration every two years; Hazard Analysis and risk-based preventive controls (Section 103) that require all FDA-regulated plants to establish Hazard Analysis Critical Control Point (HACCP) programs; targeting of inspection resources for domestic facilities, foreign facilities, and ports of entry (Section 201) that will expand FDA's abilities to inspect both domestic and foreign facilities; mandatory recall authority (Section 206) that provides FDA with the authority to stop distribution of foods and notify persons to whom the product has been distributed if a food company does not voluntarily recall product that is shown to be adulterated and poses a health hazard; and inspection of foreign food facilities (Section 306) for facilities registered with FDA as part of their requirement for export to the United States. {11}

Changes identified in the FSMA do not necessarily impact USDA-regulated meat, poultry, or egg-producing facilities. However, it is likely that USDA will review the act and consider whether sections should be adopted within the Meat Inspection Act or the Poultry Products Inspection Act [10].

USDA

While FDA has primary oversight responsibility for food, USDA, specifically the Food Safety and Inspection Service (FSIS), has responsibility for inspection, grading, and standards for meat and poultry and eggs under the Federal Meat Inspection Act (FMIA) the Poultry Products Inspection Act (PPIA) and the Egg Products Inspection Act [12]. USDA jurisdiction does not include animal by-products; however, it does include dairy products. The Agricultural Marketing Service (AMS) is responsible for fresh fruits and vegetables and grading of fresh fruits and vegetables, as well as providing audits for animal welfare.

In the past, FSIS took a very active role in sanitation; however, it was more of a command and control approach. Many plants would rely on the FSIS inspector to conduct a pre-op inspection of the plant while they followed behind and cleaned up anything the inspector found. This was referred to as the "bucket brigade," and the plant waited for "their inspector" to tell them what needed re-cleaning or corrective action. Everything was fine if "their inspector" said it was. The food industry approach has changed considerably, and plants now recognize that the inspector is not "theirs" but is a government employee, and that the plant must take responsibility for sanitation. This became more evident after January 1997.

Sanitation Standard Operating Procedures (SSOPs)

In January 1997, FSIS began initial evaluation of mandatory plant-designed Sanitation Standard Operating Procedures (SSOPs). The rationale for these rules was recognition of foodborne illness as a significant health problem in the United States. In an effort to address the statistically growing problem, the agency employed a science-based strategy to reduce the risk associated with pathogen contamination of meat and poultry products. The reliance is on prevention and the expectation that each establishment will take responsibility for sanitation and that agency personnel will monitor implementation and effectiveness of the procedures.

Part 416 of the regulations are reprinted in detail below [2]. They explain the requirement for plants to prepare written SSOPs describing the development of daily sanitation procedures, implementation, maintenance, corrective action, documentation, and agency verification. Included in the presentation of the rules are author comments to provide industry recommendations when preparing and implementing SSOPs.

416.11 General Rules

Each official establishment shall develop, implement and maintain written standard operating procedures for sanitation (Sanitation SOP's) in accordance with the requirements of this section.

Author's Comment: The original documents for review by regulatory personnel must be in English. Additional copies may be made for plant use in other languages.

416.12 Development of Sanitation SOP's

(a) The Sanitation SOP's shall describe all procedures an official will conduct daily, before and during operations, sufficient to prevent direct contamination or adulteration of product(s).

Author's Comment: It is not up to the inspector to decide what format is acceptable provided it meets regulatory requirements. There are no requirements for plans to describe in detail the mixing of chemical cleaners or the application of sanitizers. Application of chemicals must comply with the Proprietary Substances and Nonfood Compounds listing. The rule does not prescribe how cleaning will be done provided procedures can be validated as effective.

(b) The Sanitation SOP's shall be signed and dated by the individual with overall authority on-site or a higher level official of the establishment. This signature shall signify that the establishment will implement the Sanitation SOP's as specified and will maintain the Sanitation SOP's in accordance with the requirements of this part. The Sanitation SOP's shall be signed and dated upon initially implementing the Sanitation SOP's and upon any modification to the Sanitation SOP's.

Author's Comment: The signatory does not have to be an individual at the plant (i.e., can be a corporate employee), and does not have to be someone on the Grant of Inspection. Date and number each page of the SSOP.

(c) Procedures in the Sanitation SOP's that are to be conducted prior to operations shall be identified as such, and shall address, at a minimum, the cleaning of food contact surfaces of facilities, equipment and utensils.

(d) The Sanitation SOP's shall specify the frequency with which each procedure in the Sanitation SOP's is to be conducted and identify the establishment employee (s) responsible for the implementation and maintenance of such procedure (s).

Author's Comment: The plant will be required to provide data, such as a microbiological study, to justify extended clean-up, or cleaning less than every 24 hours.

Identification of employees can be by title; it does not have to be by name.

416.13 Implementation of SOP's

(a) Each official establishment shall conduct the pre-operational procedures in the Sanitation SOP's before the start of operations.

Author's Comment: The plant should assume release authority for the plant after the pre-op. It is best to inform the FSIS of the procedure for release so that they know when they can conduct their pre-op inspection.

(b) Each official establishment shall conduct all other procedures in the Sanitation SOP's at frequencies specified.
(c) Each official establishment shall monitor daily the implementation of the procedures in the Sanitation SOP's.

416.14 Maintenance of Sanitation SOP's

Each official establishment shall routinely evaluate the effectiveness of the Sanitation SOP's and the procedures therein in preventing direct contamination or adulteration of product(s) and shall revise both as necessary to keep them effective and current with respect to changes in facilities, equipment, utensils, operations or personnel.

Author's Comment: It is a good idea to maintain a page with the history of all changes made to the SSOP.

416.15 Corrective Actions

(a) Each official establishment shall take appropriate corrective action (s) when either the establishment or FSIS determines that the establishment's Sanitation SOP's or the procedures specified therein, or the implementation or maintenance of the Sanitation SOP's, may have failed to prevent the direct contamination or adulteration of product(s).

(b) Corrective actions includes procedures to insure appropriate disposition of product (s) that may be contaminated, restore sanitary conditions and prevent recurrence of direct contamination or adulteration of product(s), including appropriate reevaluation and modification of the Sanitation SOP's and the procedures specified therein.

416.16 Recordkeeping Requirements

(a) Each official establishment shall maintain daily records sufficient to document the implementation and monitoring of the Sanitation SOP's and any corrective actions taken. The establishment employee(s) specified in the Sanitation SOP's as being responsible for the implementation and monitoring of the procedure(s) specified in the Sanitation SOP's shall authenticate these records with his or her initials and the date.

Author's Comment: Prepare the inspection reports in "real time" during the inspection, not after completing the inspection. Do not rely on memory to complete the report. The report is to be prepared with a pen. If an error is made, make one mark through the error, write the correction in, and initial the correction. Do not use correction fluid to make changes.

(b) Records required by this part may be maintained on computers provided the establishment implements appropriate controls to ensure the integrity of the electronic data.

(c) Records required by this part shall be maintained for at least 6 months and made accessible and available to FSIS. All such records shall be maintained at the official establishment for 48 hours following completion, after which they may be maintained off-site provided such records can be made available to FSIS within 24 hours of request.

Author's Comment: Inspection personnel are instructed not to make copies of SSOPs or related documents. As a means of indicating that these documents are possible trade secrets, they should be marked "Confidential Commercial Information" and treated that way by the plant.

416.17 Agency Verification

FSIS shall verify the adequacy and effectiveness of the Sanitation SOP's and the procedures spelled within through review of the SSOP and review of the daily records, direct observations of implementation and direct observation of testing.

In summary, the SSOP model must detail procedures conducted daily before operation and during operation to ensure that direct product contact surfaces are cleaned to prevent direct product contamination. Surfaces such as walls and ceilings are not viewed as direct contact surfaces by the agency but should not contribute to the insanitary conditions of direct product contact surfaces. The SSOP will identify the frequency with which activities are to be conducted (Table 1.1) and will be signed by an official with overall authority (and dated) on initiation or modification and identify those responsible for implementation and daily activity. They will include procedures conducted at Pre-op, or prior to the start of operations following cleaning. The SSOP will identify records to be maintained to document implementation, monitoring, and corrective action and the individuals responsible for these activities. These records will include the appropriate disposition of product, identification of measures to ensure restoration of sanitary conditions, and prevention of reoccurrence of contamination. The establishment must routinely assess and adjust their SSOP, and the changes will be documented and subject to FSIS review.

The procedures will provide a description of cleaning process (i.e., dry pickup, rinse, disassemble, clean, and rinse), inspection steps, and sanitizing of clean equipment. They will identify the methods used by the plant for monitoring (i.e., sensory, ATP, micro) along with a sample of the inspection report used at Pre-op (Table 1.2). It should be noted that use of ATP devices or microbiological testing of product contact surfaces is not required, and it is left to the plant's discretion to determine whether they will be included in the SSOP. Operational monitoring will include equipment and utensil cleaning during and between shifts, employee hygiene, product handling in raw and cooked areas, pest control, condensation, or other factors that might influence production of product that is not adulterated. A sample of the inspection report used for Operational Sanitation monitoring will be included in the SSOP (Table 1.3). FSIS will verify compliance, from tasks to review plans, to failures from NRs (Noncompliance Reports) to withholding inspection.

Table 1.1 Sanitation Standard Operating Procedure

Cleo's Foods *Sanitation Standard Operating Procedure*		
Address: 6 Elliot Drive Hannibal, USA	Phone: (123) 456-7890	Est. No.: 6041
Cleo's Foods Management Structure: Cleo Katt—President Michael John—Quality Assurance Director Buckey Katt—Sanitation Manager		
The sanitation manager is responsible for implementation of the SSOPs. The QA Manager is responsible for the development and upkeep of the SSOPs along with the Sanitation Manager. The QA Manager is responsible for implementation of daily monitoring, record keeping, and documentation of corrective action in accordance with this protocol. All records, documents, or checklists related to this SSOP will be maintained on file and made available to FSIS personnel.		

Sanitation SOP for Establishment 6041
 	
Pre-operational Sanitation—Equipment and Facility Cleaning
 	
All direct product contact equipment will be cleaned and sanitized prior to the start of production on a daily basis.
 	
General Equipment Cleaning:
 	
  1. Equipment will be disassembled as needed prior to cleaning and parts are placed in tubs or on racks for cleaning.
 	
  2. Product debris is removed.
 	
  3. Equipment and parts are rinsed with clear water to remove the remaining debris.
 	
  4. Chemical cleaner is applied to parts and equipment, and they are cleaned according to the manufacturer's instructions.
 	
  5. Equipment and parts are rinsed with clear potable water.
 	
  6. Parts are sanitized with approved sanitizer.
 	
  7. Equipment is reassembled and re-sanitized.		

Table 1.1 (*Continued*)   Sanitation Standard Operating Procedure

Cleo's Foods *Sanitation Standard Operating Procedure*

Monitoring, Record Keeping, and Documentation:

The QA Manager will designate a QA individual to conduct daily organoleptic inspection of direct product contact equipment following implementation of the cleaning and sanitizing process and prior to operations. All findings will be documented on Cleo's Foods Pre-operational Sanitation Inspection Report. If the findings are acceptable, this box will be checked on the form.

Unacceptable findings along with Corrective Action results will also be recorded on the form.

Corrective Action:

Any unacceptable findings on product contact equipment will result in re-cleaning and re-sanitizing of the equipment. Re-inspection will be conducted by the QA individual along with the Sanitation Manager, and the employee responsible for cleaning the specific piece of equipment. Retraining will be conducted as needed. All results will be recorded on the Cleo's Foods Pre-operational Sanitation Inspection Report.

Operational Sanitation:

All food manufacturing will be conducted under sanitary conditions to prevent adulteration of food products.

1. Employees will follow Good Manufacturing Practices
2. Product Handling
3. Equipment
4. Temperatures
5. Pest Control
6. Condensation

Monitoring, Record Keeping, and Documentation:

The QA Manager will designate a QA individual, along with a production lead, to conduct daily operational inspection to ensure that employee GMPs and sanitary conditions are maintained. All findings will be documented on Cleo's Foods Operational Sanitation Inspection Report. If findings are acceptable, this box will be checked on the form. Unacceptable findings along with Corrective Action results will also be recorded on the form.

Continued

Table 1.1 (*Continued*)   Sanitation Standard Operating Procedure

Cleo's Foods *Sanitation Standard Operating Procedure*
Corrective Action: Any unacceptable findings will require immediate corrective action, including re-inspection of in-process materials or finished product to ensure that no product adulteration has occurred. Retraining of employees will be conducted as needed. All results will be recorded on the Cleo's Foods Operational Sanitation Inspection Report.
 Signed: _____ Dated: _____ Page 2 Confidential Commercial Information

FSIS will take action if findings indicate that product contact surfaces are not clean as defined in the act. Regulatory enforcement actions include preparation of an NR, tagging of equipment, product rejection/retention, a warning letter, suspension, and withdrawal. It is recommended that plant management meet with FSIS personnel on a weekly basis to review NR findings and discuss any changes made to the SSOPs to improve sanitary conditions. Regulatory Control Action is covered in greater detail later in Chapter 2 on Regulatory Inspection.

One of the benefits or the new regulations is that they are less prescriptive than those in the past. Though there are specific requirements that the plant maintain clean conditions, the means by which the plant achieves cleanliness is up to the plant. Clean is defined in the guidelines as "free of any soil, tissue, debris, chemical or other injurious substance that could contaminate a meat or poultry food product." The plant may also maintain more specific procedures to identify the cleaning of noncontact environment and equipment as well as auxiliary areas of the plant such as restrooms, locker rooms, or break rooms.

Sanitation Performance Standards (SPSs)

In an effort to supplement SSOP requirements, FSIS issued Directive 5000.1 (this replaced Directive 11000.1) to cover those areas not included in SSOPs, such as indirect contact areas, pest control, and water quality). It provides direction to FSIS field personnel on verification of facility compliance with total plant sanitation requirements. It provides a comprehensive resource for agency Consumer Safety Inspectors (CSIs) who conduct daily in-plant inspection and Enforcement, Investigation, Analysis Officers (EIAOs). Both the SSOPs and SPSs are less prescriptive than past regulations and allow plants the flexibility for innovation and technology to ensure sanitary operating conditions. However, USDA [8] still requires plants to implement effective programs to prevent unsanitary conditions that can

Table 1.2 Pre-operational Sanitation Inspection Report

Cleo's Foods *Pre-operational Sanitation Inspection Report*							
Inspector:				Date:			
Area	*Finding:*	*Acceptable*	*Unacceptable*	*Corrected*	*Re-inspected*	*Passed*	*Comments:*
Grinding							
Mixing							
Stuffing							
Cooking							
Chilling							
Slicing							
Packaging							
Packing							
Confidential Commercial Information							

Table 1.3 Operational Sanitation Inspection Report

		Acceptable	Unacceptable	Corrected	Re-inspected	Passed	
Cleo's Foods *Operational Sanitation Inspection Report*							
Inspector:				*Date:*			
Area:	*Finding:*						*Comments:*
GMPs							
Product Handling							
Equipment							
Temperatures							
Pest Control							
Condensation							
Other							
Confidential Commercial Information							

lead to product adulteration. Part 416.1 indicates that "each official establishment must be operated and maintained in a manner sufficient to prevent the creation of insanitary conditions and to ensure that product is not adulterated." Plants should develop SOPs to address each of the sections in the SPS, and an example is presented in Table 1.4. Each section of the SPSs identified below [3].

Grounds and Pest Control 416.2 (a)

The grounds about an establishment must be maintained to prevent conditions that could lead to insanitary conditions, adulteration of product, or interference with inspection by FSIS personnel. Establishments must have in place a pest management program to prevent the harborage and breeding of pests on the grounds and within establishment facilities. Pest control substances used must be safe and effective under the conditions of use and not be applied or stored in a manner that will result in the adulteration of product.

Construction 416.2 (b)

(1) Establishment buildings, including their structures, rooms, and compartments must be of sound construction, kept in good repair, and be of sufficient size to allow for processing, handling, and storage of product in a manner that does not result in product adulteration or the creation of insanitary conditions.

(2) Walls, floors, and ceilings within establishments must be built of durable materials impervious to moisture and be cleaned and sanitized as necessary to prevent adulteration of product.

(3) Walls, floors, ceilings, doors, windows, and other outside openings must be constructed and maintained to prevent the entrance of vermin, such as flies, rats, and mice.

(4) Rooms or compartments in which edible product is processed, handled, or stored must be separate and distinct from rooms or compartments in which inedible product is processed, handled, or stored, to the extent necessary to prevent product adulteration and the creation of insanitary conditions.

Lighting 416.2 (c)

Lighting of good quality and sufficient intensity to ensure that sanitary conditions are maintained and that product is not adulterated must be provided in areas where food is processed, handled, stored, or examined; where equipment and utensils are cleaned; and in hand-washing areas, dressing and locker rooms, and toilets.

Table 1.4   Sanitation Performance Standard Operating Procedure

Cleo's Foods *Sanitation Performance Standard Operating Procedure*		
Procedure Number: 1	Version: 2 Dated: 05-13-05	Replaces Version: 1 Dated: 06-26-04
Procedure: Good Manufacturing Practices (GMPs)		Pages: 3

1. Objective: The purpose of this procedure is to provide our employees and visitors with clear guidelines on expectations for sanitary personal practices during operations when food ingredients or finished products are exposed.

2. Responsibility:

    a. Plant management is responsible for enforcement and implementation of this procedure.

    b. Plant employees and visitors are responsible for compliance with this procedure.

    c. Human Resources and Quality Assurance are responsible for providing initial training to new employees and refresher training to all employees every six (6) months.

3. Monitoring, Documentation, and Corrective Action:

    a. Prior to the start of daily operations and prior to the start of the second shift, production leads and Quality Assurance will monitor employees and visitors for conformance to GMP requirements.

    b. Additionally, once per shift during daily operations, Quality Assurance will monitor employees and visitors for continued GMP conformance.

    c. Findings will be documented on the Cleo's Foods GMP Monitoring Checklist.

    d. Corrective actions will be taken for any nonconformance, including:

        i. Correction of the deviation and retraining of the employee or visitor.

        ii. Evaluation of the ingredients or products present for indication of adulteration.

        iii. Disciplinary action up to and including termination for repeated nonconformance; removal of visitors from production areas.

4. GMP Requirements:

    a. **Smocks:** Approved smocks, provided by Cleo's Foods, will be worn in the production and warehouse areas of the plant. Smocks must be removed before going outside, and entering restrooms.

Table 1.4 (*Continued*)   Sanitation Performance Standard Operating Procedure

Cleo's Foods *Sanitation Performance Standard Operating Procedure*
b. **Hairnets:** Approved hairnets, provided by Cleo's Foods, will be worn in the production and warehouse areas of the plant. All hair must be covered. Hairnets must be removed before going outside or entering restrooms.
c. **Beard Nets:** As needed, approved beard nets, provided by Cleo's Foods, will be worn in the production and warehouse areas of the plant. All facial hair must be covered. Beard nets must be removed before going outside and entering the restrooms.
d. **Food and Drinks:** Food and drinks are not to be eaten, except in approved areas (the lunchroom). All refuse must be properly disposed of. Gum or candy is not permitted in the production area.
e. **Jewelry:** No jewelry is to be worn in the plant at anytime. This includes watches, rings, earrings or exposed piercing, bracelets, and necklaces. Medic Alert jewelry is allowed.
f. **Product Contamination Prevention:** No glass is allowed in production areas. This includes glass containers, meters, tools, or utensils that have glass as a part.
g. **Smoking and Tobacco:** Smoking or the use of any tobacco product (chew, snuff) is not allowed anywhere in the plant. Smoking is only allowed in the designated areas. Do not throw cigarette butts on the ground. The butts must be properly disposed of.
h. **Spitting:** Spitting is prohibited in all areas.
i. **Outside Doors:** All outside doors must remain closed at all times and will not be propped opened by contractors.
j. **Hand Washing:** All employees and visitors must wash and sanitize their hands after using the restroom and before entering any production area.
Regulatory References: 9CFR 416.5, 21CFR 110.10
Confidential Commercial Information

Ventilation 416.2 (d)

Ventilation adequate to control odors, vapors, and condensation to the extent necessary to prevent adulteration of product and the creation of insanitary conditions must be provided.

Plumbing and Sewage 416.2 (e) and (f)

Plumbing systems must be installed and maintained to:

(1) Carry sufficient quantities of water to required locations throughout the establishment;
(2) Properly convey sewage and liquid disposable waste from the establishment;
(3) Prevent adulteration of product, water supplies, equipment, or utensils, and maintain sanitary conditions throughout the establishment;
(4) Provide adequate floor drainage in all areas where floors are subject to flooding-type cleaning or where normal operations release or discharge water or other liquid waste on the floor;
(5) Prevent back-flow conditions in and cross-connection between piping systems that discharge waste water or sewage and piping systems that carry water for product manufacturing; and
(6) Prevent the backup of sewer gases.

Sewage must be disposed into a sewage system separate from all other drainage lines or disposed of through other means sufficient to prevent backup of sewage into areas where product is processed, handled, or stored. When the sewage disposal system is a private system requiring approval by a State or local health authority, the establishment must furnish FSIS with the letter of approval from that authority upon request.

Water Supply and Reuse 416.2 (g)

(1) A supply of running water that complies with the National Primary Drinking Water regulations (40 CFR Part 141), at a suitable temperature and under pressure as needed, must be provided in all areas where required (for processing product, for cleaning rooms and equipment, utensils, and packaging materials, for employee sanitary facilities, etc.). If an establishment uses a municipal water supply, it must make available to FSIS, upon request, a water report, issued under the authority of the State or local health agency, certifying or attesting to the potability of the water supply. If an establishment uses a private well for its water

supply, it must make available to FSIS, upon request, documentation certifying the potability of the water supply, that has been renewed at least semi-annually.

(2) Water, ice, and solutions (such as brine, liquid smoke, or propylene glycol) used to chill or cook ready-to-eat product may be reused for the same purpose, provided that they are maintained free of pathogenic organisms and fecal Coliform organisms and that other physical, chemical, and microbiological contamination have been reduced to prevent adulteration of product.

(3) Water, ice, and solutions used to chill or wash raw product may be reused for the same purpose provided that measures are taken to reduce physical, chemical, and microbiological contamination so as to prevent contamination or adulteration of product. Reuse water which has come into contact with raw product may not be used on ready-to-eat product.

(4) Reconditioned water that has never contained human waste and that has been treated by an onsite advanced wastewater treatment facility may be used on raw product, except in product formulation, and throughout the facility in edible and inedible production areas, provided that measures are taken to ensure that this water meets the criteria prescribed in paragraph (g) (1) of this section. Product, facilities, equipment, and utensils coming in contact with this water must undergo a separate final rinse with non-reconditioned water that meets the criteria prescribed in paragraph (g)(1) of this section.

(5) Any water that has never contained human waste and that is free of pathogenic organisms may be used in edible and inedible product areas, provided it does not contact edible product. For example, such reuse water may be used to move heavy solids, flush the bottom of open evisceration troughs, or to wash antemortem areas, livestock pens, trucks, poultry cages, picker aprons, picking room floors, and similar areas within the establishment.

(6) Water that does not meet the use conditions of paragraphs (g)(1) through (g)(5) of this section may not be used in areas where edible product is handled or prepared or in any manner that would allow it to adulterate edible product or create insanitary conditions.

Dressing Room/Lavatory 416.2 (h)

(1) Dressing rooms, toilet rooms, and urinals must be sufficient in number, ample in size, conveniently located, and maintained in a sanitary condition and in good repair at all times to ensure cleanliness of all persons handling any product. They must be

separate from the rooms and compartments in which products are processed, stored, or handled.

(2) Lavatories with running hot and cold water, soap, and towels, must be placed in or near toilet and urinal rooms and at such other places in the establishment as necessary to ensure cleanliness of all persons handling any product.

(3) Refuse receptacles must be constructed and maintained in a manner that protects against the creation of insanitary conditions and the adulteration of product.

Equipment and Utensils 416.3

(a) Equipment and utensils used for processing or otherwise handling edible product or ingredients must be of such material and construction to facilitate thorough cleaning and to ensure that their use will not cause the adulteration of product during processing, handling, or storage. Equipment and utensils must be maintained in sanitary condition so as not to adulterate product.

(b) Equipment and utensils must not be constructed, located, or operated in a manner that prevents FSIS personnel from inspecting the equipment or utensils to determine whether they are in sanitary condition.

(c) Receptacles used for storing inedible material must be of such material and construction that their use will not result in the adulteration of any edible product or in the creation of insanitary conditions. Such receptacles must not be used for storing any edible product and must bear conspicuous and distinctive marking to identify permitted uses.

Sanitary Operations 416.4

(a) All food-contact surfaces, including food-contact surfaces of utensils and equipment, must be cleaned and sanitized as frequently as necessary to prevent the creation of insanitary conditions or the adulteration of product.

(b) Non-food-contact surfaces of facilities, equipment, and utensils used in the operation of the establishment must be cleaned and sanitized as frequently as necessary to prevent the creation of insanitary conditions or the adulteration of product.

(c) Cleaning compounds, sanitizing agents, processing aids, and other chemicals used by an establishment must be safe and effective under the conditions of use. Such chemicals must be used, handled, and stored in a manner that will not adulterate product or create insanitary conditions. Documentation substantiating the safety of a chemical's use in a food processing environment must be available to FSIS inspection personnel for review.

(d) Product must be protected from adulteration during processing, handling, storage, loading, and unloading at and during transportation from official establishments.

Employee Hygiene 416.5

(a) Cleanliness. All persons working in contact with product, food-contact surfaces, and product-packaging materials must adhere to hygienic practices while on duty to prevent adulteration of product.

(b) Clothing. Aprons, frocks, and other outer clothing worn by persons who handle product must be of material that is disposable or readily cleaned. Clean garments must be worn at the start of each working day and garments must be changed during the day as often as necessary to prevent contamination or adulteration of product.

(c) Disease control. Any person who has or appears to have an infectious disease, open lesion, including boils, sores, or infected wounds, or any other abnormal source of microbial contamination must be excluded from any operations which could result in product adulteration until the condition is corrected.

Tagging Insanitary Equipment, Utensils, Rooms or Compartments 416.6

When a Program employee finds that any equipment, utensil, room, or compartment at an official establishment is insanitary or that its use could cause the adulteration of product, he will attach to it a "U.S. Rejected" tag. Equipment, utensils, rooms, or compartments so tagged cannot be used until made acceptable. Only a Program employee may remove a "U.S. Rejected" tag.

As you can see, the Sanitation Performance standards provide a comprehensive guideline for maintaining overall plant cleanliness. In Chapter 6, the application of these standards will be demonstrated through sanitary equipment and facility design. Again, it is recommended that USDA plants use the information provided in the SPSs to develop plant food safety systems and sanitation procedures to ensure sanitary conditions, prevent adulteration of product, and meet regulatory requirements. Familiarity with the FDA GMPs, USDA SSOPs, and USDA SPSs will assist plants and food companies prepare food safety systems that will both meet regulatory requirements and also go a long way toward ensuring the safety and wholesomeness of their products.

European Food Safety Authority

The European Food Safety Authority (EFSA) is an agency of the European Union (EU) established in 2002 and based in Parma, Italy. It was established as a result of a series of food safety crises in the late 1990s. It is considered key to the EU's efforts to ensure a high level of consumer food protection and to enhance confidence in the food supply. It supports member states with effective and timely risk management decisions to protect consumers and the food and feed chain [13,14].

The agency was established to provide independent scientific advice on all matters that directly or indirectly impact food or animal feed. This involves an integrated approach to food safety from farm to fork, with monitoring of compliance with expectations. It includes expectation of timely communication on existing or emerging risks related to food and feed [15,16].

Regulation 178/2002 establishes the EFSA and the general principles and requirements of food law. It lays down the procedures for the distribution of safe food to ensure the quality of food for humans or for animal feed and to ensure that no food that is unfit for consumption enters commerce. It is also intended to protect consumers from fraud and deceptive practices by producers and distributors of food. Food manufacturers must apply the appropriate regulations at all stages of the production and distribution process, from processing to transportation and distribution [17].

The regulation also requires that food manufacturers have effective traceability systems for ingredients and finished products. If a manufacturer considers a food item to be potentially harmful, they must initiate a withdrawal. This identification is verified when the manufacturer conducts a risk analysis using science-based information to assess the risk of the food. The following factors are used to assess whether the food could be dangerous for human consumption:

■ The normal conditions of use
■ Information provided to the consumer (use, handling)
■ Whether there is an immediate or delayed effect
■ Whether there are cumulative toxic effects
■ Sensitivity of specific consumers

Once this is done, and the manufacturer initiates a withdrawal, they must also notify the appropriate authority and provide timely notification to consumers. If only part of a batch or lot is considered harmful for consumption, then the entire batch or lot is considered safe and subject to withdrawal. A rapid alert system, RAPEX, aids in information exchange concerning the action to restrict or withdraw a food item [17].

Regulation 852/2004 applies to food manufacturers and establishes the practices to ensure the hygiene of foodstuffs. It requires that the manufacturers be registered with the appropriate authority and that they carry out their practices in a hygienic manner. The scope of hygienic processes includes

- The food premises, including the surrounding site area
- Transport conditions
- Equipment
- Food waste
- Water supply
- Personal hygiene of food handlers
- Food
- Wrapping and packaging
- Heat treatment used to process certain foods
- Training of food workers

Food businesses must also comply with Regulation 853/2004, which provides specific hygiene rules and specific rules concerning microbiological criteria for foods as well as temperature controls and cold chain compliance. Food operators will also apply the seven principles of Hazard Analysis Critical Control Point (HACCP) as identified in Codex Alimentarius [18].

References

1. U.S. Food and Drug Administration, *Current Good Manufacturing Practices in Manufacturing: Packing and Holding Human Food*, Washington, D.C.
2. U.S. Department of Agriculture, *Sanitation Standard Operating Procedures*, Washington, D.C.
3. U.S. Department of Agriculture, *Sanitation Performance Standards*, Washington, D.C.
4. Gould, Dr. Wilbur A., *CGMP's/Food Plant Sanitation*, 2nd edition, CTI Publications, Baltimore, 1994, chaps. 2 and 3.
5. Stauffer, John E., *Quality Assurance of Food Ingredients Processing and Distribution*, 3rd edition, Food and Nutrition Press, Trumbull, 1994, chap. 6.
6. Troller, John A., *Sanitation in Food Processing*, Academic Press, New York, 1983, chap. 19.
7. Keener, Larry, *Sanitation Initiatives and Innovations, Integrating Regulations and Standards*, presented at "Equipped for Excellence," FPM Expo, Las Vegas, September 26, 2005.
8. U.S. Department of Agriculture, *Supervisory Guidelines for the Sanitation Performance Standards*, Washington, D.C.
9. Sinclair, Upton, *The Jungle*, Signet Classic Printing, New York, 1960.
10. Sanchez, Marc, Understanding the Ancestry of the Food Safety Modernization Act, *Food Safety Magazine*, August/September 2011, pp. 24–27.
11. Food Safety Modernization Act, FDA Website, fda.gov/FoodSafety/FSMA/ucm247548.

12. Katsuyama, Allen M., *Principles of Food Processing Sanitation*, Food Processors Institute, 1993.
13. European Food Safety Authority, en.wikipedia.org/wiki/European_Food_Safety_Authority.
14. About EFSA, efsa.europa.eu/en/aboutefsa.htm.
15. European Union, Agencies and Other EU Bodies, European Food Safety Authority (EFSA), europa.eu/agencies/regulatory_agencies_bodies/policy_agencies/efsa/indez_en.html.
16. Food Safety, eurpoa.eu/legislation_summaries/food_safety/index_en.html.
17. Food and Feed Safety, europa.eu/legislation_summaries/food_safety/general_provisions/f80501_en.html.
18. Food Hygiene, europa.eu/legislation_summaries/food_safety/veterinary_provisions/f80501_en.html.

Chapter 2

Regulatory Inspection and Control Action

The rules identified in Chapter 1 demonstrate the statutory authority of both the FDA and the USDA to regulate food manufacturing and distribution. Both agencies also have the authority to conduct inspections in the plant. The USDA Food Safety and Inspection Service (FSIS) typically has inspectors in the plant, referred to as Consumer Safety Inspectors (CSIs), on a regular basis are supported by their direct supervisor, the Front Line Supervisor, and can be supplemented by Enforcement, Inspection, Analysis Officers (EIAOs), who are typically brought in to conduct in-depth verification of a plant food safety system. The FDA, however, does not have inspectors on site in food manufacturing plants. They rely on inspectors to go into facilities on a routine or directed basis to conduct inspections. This chapter identifies the rights of inspectors to conduct inspections and their means of control when violations are suspected or are found. It also provides food plants with the basis for preparing an internal protocol for handling regulatory inspections. It will also provide food manufacturers with the basis for preparing a recall program in the event inspection findings lead to the need to pull product from distribution.

Inspection

The Federal Food, Drug and Cosmetic Act authorizes FDA inspectors, upon presentation of credentials and notice to the owner or operator, to enter at reasonable times for the purpose of inspection, any factory warehouse or establishment where foods are manufactured, processed, packaged, or held for introduction into interstate commerce, or after they have been shipped interstate, and to enter any vehicle used to transport or to hold any such food. Regulators have legal authority to access food plants during "reasonable hours" for purposes of inspection. Reasonable hours may be interpreted as any time of day when the plant is staffed and in production [1]. The plant cannot deny access during these reasonable times, and it would not be wise to do so. As long as the entry and inspection is conducted within the reasonable times, limits and manners specified, as a general policy, the company should permit an entry and inspection to proceed without a warrant, keeping in mind that you may reserve the right to require a warrant [3]. Within the limits established by law and consistent with protection proprietary information and trade secrets, it should be company policy to cooperate fully and courteously within reason with federal, state, and local officials engaged in inspections to confirm compliance with applicable law and regulation. Again, provided the inspection occurs at reasonable times and within reasonable limits and in a reasonable manner, his authority includes the plant and any vehicle used to transport food and all pertinent equipment, finished and unfinished materials and ingredients, and containers and labeling.

The plant should have an established procedure for handling regulatory inspections. Regulatory personnel may be required to provide official credentials, unless they are known personally, and may be requested to sign in for plant security purposes. Inspectors of the FDA must also give written notice to the manager of the inspected plant, or to the person in charge in his absence, before inspection can proceed. It is a good idea to do everything to accommodate the inspectors and not keep them waiting. Begin with an introductory meeting with the inspector to present the plant management team. Determine ahead of time who will be in the meeting and who will accompany the inspector through the plant, and assemble them as soon as possible. Preferably the group will be multi-functional, production, maintenance, quality, and sanitation. At this time, the inspector should present the Notice of Inspection (Form 482) (Table 2.1) [1]. Regulatory inspection may be routine or for cause, meaning that they may have a complaint that they are investigating or the inspection may be because of a regulatory or food safety violation that resulted in a recall. During the introductory meeting, inquire as to the reason for the visit, whether it is routine or for cause. Make a note of inspector's name and the agency he or she represents. Some agencies are not authorized to inspect any given facility, yet some will attempt to do so despite a lack of authority. If you are in doubt regarding an inspector's authority, ask politely that he explain his authority to you and/or contact your representative trade association or legal counsel for advice.

If the inspectors take a plant tour, it is a good idea to have one more person with the inspectors than the number of inspectors, and it's always best to have a cross-functional group so that questions can be answered. Do not stop operations during the inspection; continue plant operations as normal. When responding to the inspector, answer the questions honestly and succinctly, but do not volunteer more information than is requested. If plant records are requested, provide only those that are requested and only upon request. There are records that inspectors

Table 2.1 Form FDA-482

DEPARTMENT OF HEALTH AND HUMAN SERVICES PUBLIC HEALTH SERVICE FOOD AND DRUG ADMINISTRATION		1. DISTRICT ADDRESS & PHONE NO. Rm 508 Federal Office Building 30 U.N. Plaza San Francisco, CA 94102 (415) 556-2062	
TO	2. NAME AND TITLE OF INDIVIDUAL Robert K. Thompson, Plant Manager	3. DATE 5-15-85	
	4. FIRM NAME Garden City Nut Shellers	5. HOUR	8:30 a.m.
	6. NUMBER AND STREET 2704 Sellers Ave		p.m.
	7. CITY AND STATE & ZIP CODE San Jose, CA 95131	8. PHONE # & AREA CODE (408)123-4567	

Notice of Inspection is hereby given pursuant to Section 704(a)(1) of the Federal Food, Drug, and Cosmetic Act [21 U.S.C. 374(1)]1 and/or Part F or G, Title III of the Public Health Service Act [42 U.S.C. 262-264]2

9. SIGNATURE (Food and Drug Administration Employee(s))	10. TYPE OR PRINT NAME AND TITLE (FDA Employee(s))
	Sidney H. Rogers, Investigator

[1]Applicable to portions of Section 704 and other Sections of the Federal Food, Drug, and Cosmetic Act [21 U.S.C. 374] are quoted below:

Sec. 704. (a)(1) For purposes of enforcement of this Act, officers or employees duly designated by the Secretary, upon presenting appropriate credentials and a written notice to the owner, operator, or agent in charge, are authorized (A) to enter, at reasonable times, any

[2]Applicable sections of Parts F and G of Title III Public Health Service Act[42 U.S.C. 262-264] are quoted below:

Part F—Licensing—Biological Products and Clinical Laboratories and ******

Sec. 351(c) "Any officer, agent, or employee of the Department of Health & Human Services, authorized by the Secretary for the purpose, may during all reasonable hours enter and

Continued

Table 2.1 (*Continued*)   Form FDA-482

factory, warehouse, or establishment in which food, drugs, devices, or cosmetics are manufactured, processed, packed, or held, for introduction into interstate commerce or after such introduction, or to enter any vehicle being used to transport or hold such food, drugs, devices, or cosmetics in interstate commerce; and (B) to inspect, at reasonable times and within reasonable limits and in a reasonable manner, such factory, warehouse, establishment, or vehicle and all pertinent equipment, finished and unfinished materials, containers and labeling therein. In the case of any factory, warehouse, establishment, or consulting laboratory in which prescription drugs, nonprescription drugs intended for human use, or restricted devices are manufactured, processed, packed, or held, the inspection shall extend to all things therein (including records, files, papers, processes, controls, and facilities) bearing on whether prescription drugs, nonprescription drugs intended for human use, or restricted devices which are adulterated or misbranded within the meaning of this Act, or which may not be manufactured, introduced into interstate commerce, or sold, or offered for sale by reason of any provision of this Act, have been or are being manufactured, processed, packed, transported, or held in any such place, or otherwise bearing on violation of this Act. No inspection authorized by the preceding sentence or by paragraph (3) shall extend to financial data, sales data other then shipment data, pricing

inspect any establishment for the propagation or manufacture and preparation of any virus, serum, toxin, antitoxin, vaccine, blood, blood component or derivative, allergenic product or other product aforesaid for sale, barter, or exchange in the District of Columbia, or to be sent, carried, or brought from any State or possession into any other State or possession or into any foreign country, or from any foreign country into any State or possession."

Part F - ****** Control of Radiation.

Sec. 360 A (a) "If the Secretary finds for good cause that the methods, tests, or programs related to electronic product radiation safety in a particular factory, warehouse, or establishment in which electronic products are manufactured or held, may not be adequate or reliable, officers or employees duly designated by the Secretary, upon presenting appropriate credentials and a written notice to the owner, operator, or agent in charge, are thereafter authorized (1) to enter, at reasonable times any area in such factory, warehouse, or establishment in which the manufacturer's tests (or testing programs) required by section 358(h) are carried out, and (2) to inspect, at reasonable times and within reasonable limits and in a reasonable manner, the facilities and procedures within such area which are related to electronic product radiation safety. Each such inspection shall be commenced and completed with reasonable promptness. In addition to other grounds upon which good cause may be found for

Table 2.1 (*Continued*)   Form FDA-482

data, personnel data (other than data as to qualifications of technical and professional personnel performing functions subject to this Act), and research data (other than data, relating to new drugs, antibiotic drugs and devices and, subject to reporting and inspection under regulations lawfully issued pursuant to section 505(i) or (k), section 507(d) or (g), section 519, or 520(g), and data relating to other drugs or devices which in the case of a new drug would be subject to reporting or inspection under lawful regulations issued pursuant to section 505(j) of the title). A separate notice shall be given for each such inspection, but a notice shall not be required for each entry made during the period covered by the inspection. Each such inspection shall be commenced and completed with reasonable promptness.

Sec. 704(e) Every person required under section 519 or 520(g) to maintain records and every person who is in charge or custody of such records shall, upon request of an officer or employee designated by the Secretary, permit such officer or employee at all reasonable times to have access to and to copy and verify, such records.

Sec. 704(f)(1) A person accredited under section 523 to review reports under section 510(k) and make recommendations of initial classifications of devices to the Secretary shall maintain records documenting the training qualifications of the person and the

purposes of this subsection, good cause will be considered to exist in any case where the manufacturer has introduced into commerce any electronic product which does not comply with an applicable standard prescribed under this subpart and with respect to which no exemption from the notification requirements has been granted by the Secretary under section 359(a)(2) or 359(e)."

(b) "Every manufacturer of electronic products shall establish and maintain such records (including testing records), make such reports, and provide such information, as the Secretary may reasonably require to enable him to determine whether such manufacturer has acted or is acting in compliance with this subpart and standards prescribed pursuant to this subpart and shall, upon request of an officer or employee duly designated by the Secretary, permit such officer or employee to inspect appropriate books, papers, records, and documents relevant to deter-mining whether such manufacturer has acted or is acting in compliance with standards prescribed pursuant to section 359(a)."

****

Continued

Table 2.1 (*Continued*)   Form FDA-482

employees of the person for handling confidential information, the compensation arrangements made by the person, and the procedures used by the person to identify and avoid conflicts of interest. Upon the request of an officer or employee designated by the Secretary, the person shall permit the officer or employee, at all reasonable times, to have access to, to copy, and to verify, the records.	
Section 512 (l)(1) In the case of any new animal drug for which an approval of an application filed pursuant to subsection (b) is in effect, the applicant shall establish and maintain such records, and make such reports to the Secretary, of data relating to experience and other data or information, received or otherwise obtained by such applicant with respect to such drug, or with respect to animal feeds bearing or containing such drug, as the Secretary may by general regulation, or by order with respect to such application, prescribe on the basis of a finding that such records and reports are necessary in order to enable the Secretary to determine, or facilitate a determination, whether there is or may be ground for invoking subsection (e) or subsection (m)(4) of this section. Such regulation or order shall provide, where the Secretary deems it to be appropriate, for the examination, upon request, by the persons to whom such regulation or order is applicable, of similar in-formation received or otherwise obtained by the Secretary.	

Table 2.1 (*Continued*)   Form FDA-482

(2) Every person required under this subsection to maintain records, and every person in charge or custody thereof, shall, upon request of an officer or employee designated by the Secretary, permit such officer or employee at all reasonable times to have access to and copy and verify such records.	
FORM FDA 482(9/00) *PREVIOUS EDITION* *NOTICE OF INSPECTION* *IS OBSOLETE*	
(Reverse of Form FDA 482)	
"The Secretary may by regulation (1) require dealers and distributors of electronic products, to which there are applicable standards prescribed under this subpart and the retail prices of which is not less than $50, to furnish manufacturers of such products such information as may be necessary to identify and locate, for purposes of section 359, the first purchasers of such products for purposes other than resale, and (2) require manufacturers to preserve such information. Any regulation establishing a requirement pursuant to clause (1) of the preceding sentence shall (A) authorize such dealers and distributors to elect, in lieu of immediately furnishing such information to the manufacturer to hold and preserve such information until advised by the manufacturer or Secretary that such information is needed by the manufacturer for purposes of section 359, and (B) provide that the dealer or distributor shall, upon making such election, give prompt notice of such election (together with information	Sec. 360 B.(a) It shall be unlawful— (1) *** (2) *** (3) "for any person to fail or to refuse to establish or maintain records required by this subpart or to permit access by the Secretary or any of his duly authorized representatives to, or the copying of, such records, or to permit entry or inspection, as required or pursuant to section 360A." *** Part G—Quarantine and Inspection Sec. 361(a) "The Surgeon General, with the approval of the Secretary is authorized to make and enforce such regulations as in his judgment are necessary to prevent the introduction, transmission, or spread of communicable diseases from foreign countries into the States or possessions, or from one State or possession into any other State or possession. For purposes of carrying out and enforcing such regulations, the Surgeon General may provide for

Continued

Table 2.1 (*Continued*)   Form FDA-482

identifying the notifier and the product) to the manufacturer and shall, when advised by the manufacturer or Secretary, of the need therefore for the purposes of Section 359, immediately furnish the manufacturer with the required information. If a dealer or distributor discontinues the dealing in or distribution of electronic products, he shall turn the information over to the manufacturer. Any manufacturer receiving information pursuant to this subsection concerning first purchasers of products for purposes other than resale shall treat it as confidential and may use it only if necessary for the purpose of notifying persons pursuant to section 359(a)."	such inspection, fumigation, disinfection, sanitation, pest extermination, destruction of animals or articles found to be so infected or contaminated as to be sources of dangerous infection to human beings, and other measures, as in his judgment may be necessary."

are allowed to view and those that they have no access to, such as consumer complaints, formula or processing data, or cleaning schedules. FDA inspectors have access to information relating to ingredient receipt, processing and packaging, and shipment of food products. They may not examine any financial data, sales data other than product shipment data, pricing data, personnel data, or research data. Do not reveal any product costing records, production yields, or profits to regulatory agents. USDA inspectors can inspect records on Child Nutrition Products regarding formulas, processing procedures, and ingredient lists only. They do not have access to documents such as personnel files or financial data [6].

FDA inspectors taking samples of finished product for analysis are required to provide the firm with a Receipt (Form 484) [1]. If the inspector takes samples for microbiological analysis, it is highly recommended that all product from the line sampled be retained from the last full cleanup, to the next full cleanup. If any of the product is already in distribution, place it on hold even if it is already at a customer location. Taking a duplicate sample is a good idea for future reference; however, consider the consequences of analyzing the duplicate sample. If the regulatory sample analyzed for pathogens yields a positive result, your split result will not offset the positive regulatory result. Product will still be considered adulterated and subject to disposal. On the other hand, if the regulatory sample tests negative for pathogens and your split tests positive, product is considered adulterated and subject to disposal.

Only OSHA inspectors have statutory authority to take photographs in the plant. FDA and USDA inspectors are not specifically authorized to take photographs within the plant or to operate a tape or other recording devices. Although use of tape or other recorders and dictation equipment is a convenient shorthand means of taking notes, the taking of photographs can disclose confidential processing methods or other trade secrets. The best company policy is to not permit the use of a camera or recorder. However, FDA inspectors may insist that they also have the legal authority to take pictures [3]. It is best if the plant prohibits photography in the facility and important that you inform the inspector of this fact and expect that they comply. If they press, you may request that they obtain a warrant; however, you must balance this with the spirit of cooperation during the inspection. If the inspector obtains a warrant permitting photos or recordings, it is best to comply, but you should take duplicate photos as well as photos of the surroundings. The best advice is that during any regulatory inspection, if photos are taken, request duplicate copies or that the plant takes duplicate photos of the same location and surrounding area.

Upon completion of the inspection and before the inspector leaves the premises, request a meeting to discuss in detail any findings that the inspector plans to record. An FDA inspector should give the owner or operator a report (Form 483) in writing setting forth any conditions or practices that he or she has observed, which in his or her judgment indicate that any food in the plant consists in whole or in part of any filthy, putrid, or decomposed substance or has been prepared, packed, or held under unsanitary conditions and therefore may have become contaminated or rendered injurious to health [1]. It is prudent to discuss the inspection with the inspector in detail. If you do not understand an item, ask about it, or if you do not agree with a particular observation, explain your position. If you have corrected an observation during the inspection, inform the inspector. Also, ask the inspector to make any appropriate changes in his or her list at this time. If you intend to correct certain observations, explain this. Even if the inspector does not amend the list, ask him or her to include your comments in the report. You want to emphasize to the inspector that you are taking all reasonable steps to manufacture clean, safe, and accurately labeled products. If a signed report is required, consider having your legal counsel review the report before signing. Although an inspector may indicate impatience with waiting for legal review, he or she cannot demand your signature if you request a legal review. If, during the review of the report discrepancies are noted, insist that the discrepancies be corrected or note the discrepancies on the report before signing. Though it may not always be required or requested by the agency, it is highly recommended that you provide a follow-up report to document any and all corrective actions. Remember, many of the regulatory reports are subject to access through the Freedom of Information Act (FOIA); thus, your response provides requestors with documentation of your intentions to provide for a sanitary environment and prevent product adulterations.

Regulatory Control Action

Following an FDA inspection, there are several possible outcomes that may occur. The most desirable outcome is that there are no findings of adverse or insanitary conditions, or "minor" recommendations for corrective action. These deficiencies should be addressed as soon as possible to avoid repeat findings on ensuing inspections. However, if there are more serious findings, the inspector may issue a **Notice of Adverse Findings** for conditions that may lead to a regulatory violation if corrective action is not taken. This is usually issued if there is indication that the firm will take prompt corrective action, and thus the violation does not require further action against the firm [2]. A **regulatory letter** may be issued when prompt voluntary correction is sought or to warn of possible regulatory or legal actions if correction is not prompt [2]. A follow-up inspection may be required to verify implementation and effectiveness of corrective actions. If the violations are severe enough, the FDA may initiate a product **seizure** to remove violative product from commerce. In this case, a complaint is filed in the District Court, and a US Marshall tags the product as Seized. If there is a history of violations, the FDA may request the court to issue an **injunction** to prevent shipment of violative product [3]. Other FDA tools include disclosure and publicity alerting consumers of problems relating to the firm's food. In any of these instances, immediate and effective action is recommended to correct any and all deficiencies to protect consumers, stop production of adulterated product, and discontinue actions that can severely damage a company's reputation. All inspection and action protocol is addressed in the FDA Regulatory Procedures Manual.

USDA has three types of enforcement actions as defined in the agency's Rules of Practice (9 CFR 500) [10]. These actions include "regulatory control action," "withholding action," and "suspension."

1. Regulatory Control Action involves the retention of product or rejection of facilities or equipment. It may include the slowing or stopping of lines or prevention of processing of specific product.
2. Withholding Action involves the refusal to allow the mark of inspection to be applied to product on some or all product in the establishment. Without approval to use the mark, product cannot be sold.
3. Suspension is the interruption in the assignment of inspection to all or part of the establishment. Without inspection assignment, the establishment cannot run.

These actions may be taken if there are findings of insanitary conditions or practices, product adulteration or misbranding, conditions that prevent agency personnel from determining product is not adulterated or misbranded, or inhumane handling or slaughtering practices. The intent of the control action is to address specific problems identified by agency personnel until they have been addressed by the establishment. Under normal conditions, the on-site CSI will issue a Noncompliance Report (NR) for deviations discovered while performing inspection tasks or non-routine inspection, especially during pre-op inspection. Deviations found during pre-op may be recorded on an NR; however, if the deviations involve establishment failure to comply with SSOP requirements or its SSOP plan, an **Individual SSOP Failure** has occurred. This is especially critical when direct product contamination occurs. In this instance, the CSI may apply a US Rejected/Retained tag to equipment, in-process materials, or finished product if they believe that the deviation affects any of these items (Figure 2.1). An **SSOP System Failure** occurs when plant control systems are failing to ensure that product is not adulterated. An indicator of this may be repeat individual failures.

No prior notification is required for withholding or suspension action when there is shipment of adulterated product; there is no written HACCP plan or SSOP;

Figure 2.1 **The U.S. Rejected/Retained Tag may only be removed by a USDA employee and may not be removed by company personnel. Failure to comply may result in further regulatory action.**

there is basic noncompliance; there are gross insanitary conditions; there is the threat of assault, intimidation, or interference with inspection; there is failure to destroy adulterated product; or inhumane slaughter is documented. Prior notification of withholding is required when there is repetitive failure with no shipment of adulterated product, there are multiple recurring noncompliances that have lead to HACCP system inadequacy, the SSOP is not properly implemented or maintained, there are repeated and recurring sanitation deficiencies, there is failure to collect and analyze generic *E. coli* samples and record results, and/or *Salmonella* performance standards are not met.

Other enforcement tools used by FSIS include the **30-Day Letter;** this is generally issued as a request for clarification of rationale for food safety systems decisions or a **Notice of Intended Enforcement (NOIE)** that is issued when serious deviations are detected in a plant food safety systems. The plant has 30 days to respond to the findings in a 30-Day Letter but only 72 hours to provide adequate corrective action response to the NOIE. Failure to respond adequately or in a timely basis may result in further enforcement action up to and including Suspension of Inspection, product seizure, or Recommendation of Product Recall.

As with FDA enforcement action, the plant is well advised to take the appropriate, documented corrective action to avoid costly and damaging USDA enforcement action. However, with a well-designed SSOP and SPS and a complete food safety systems plan, these types of situations can be prevented.

Recall

While food manufacturing plants are well advised to do everything they can to prevent the manufacturing and distribution of violative products, there are circumstances where adulterated food product may leave a facility and be subject to a recall. In these instances the company would be well advised to understand the regulatory requirements of recall and have a procedure in place to facilitate the recall process. Recall is a voluntary action to remove adulterated or misbranded product from commerce. Regulatory agencies such as USDA do not have statutory authority to require food manufacturing facilities to recall product. However, USDA may become aware of adulterated or misbranded product through test results of sampling programs, consumer complaints, or company notification [8]. At that time, the agency may recommend that the company conduct a recall. The company may refuse to act on the recommendation; however, the agency has very broad powers that include the following: withdrawal of inspection, retention or seizure of product, and media notification of the recall proposal along with the company response refusal [5]. This type of disclosure would obviously not be good publicity for the company, so unless the company has solid scientific or legal reasons not to recall product, they would be advised to comply with the recommendation of the agency.

Recall Policy

When developing a Recall Policy, each facility should begin by stating that their objective is to distribute food products that are safe, wholesome, and in compliance with all applicable laws and government regulations, and to respond in a timely manner to problems involving customer/consumer protection or customer/consumer safety. In the event that shipped product needs to be recalled or withdrawn, the company has a policy that establishes a recall planning system, the purposes of which are to protect customer/consumer safety, protect the assets of the company, comply with applicable laws and government regulations, remove unacceptable or questionable products from the market at minimum cost and inconvenience to the customer/consumer, and minimize expense to the company.

There are three major definitions used by the U.S. Food and Drug Administration to describe removal of product from distribution channels. These terms are defined in the U.S. Code of Federal Regulations (21CFR7.3) as follows:

1. Recalls are the most serious of product recoveries as they represent potential for harm to consumers. They are broken down by class depending on their severity.
   a. Class I is a situation in which there is a reasonable probability that the use of, or exposure to, a violative product will cause serious adverse health consequences or death. Examples would be presence of pathogens such as *Listeria monocytogenes* in Ready To Eat product, *Clostridium botulinum* in canned goods, or presence of *E. coli* O157:H7 in ground beef. It would also include the presence of undeclared allergens, especially if those allergens have been associated with fatalities in sensitive individuals (i.e., peanuts).
   b. Class II is a situation in which use of, or exposure to, a violative product may cause temporary or medically reversible adverse health consequences or where the probability of serious adverse health consequences is remote. An example of this may include the presence of undeclared allergens that have not been associated with fatality in sensitive individuals (i.e., wheat).
   c. Class III is a situation in which the use of, or exposure to, a violative product is not likely to cause adverse health consequences. This would likely include examples of what are considered to be "economic adulteration" (i.e., underweight, excess water etc.).
2. Market Withdrawal means removal or correction of a distributed product that involves a minor violation that would not be subject to legal action by the Food and Drug Administration or which

involves no violation. For example product that is outside of normal stock rotation practices, failure of generic microbiological testing, and quality defects that do not render the product harmful but would likely lead to customer or consumer dissatisfaction.

3. Stock Recovery means removal or correction of a product that has not been marketed nor left the direct control of the firm. For example, the product is located on premises owned by, or under the control of, the firm and no portion of the lot has been released into commerce for sale or use. This would include product in a third-party distribution warehouse in which storage space is leased by the manufacturer.

Recall Procedure

Using the definitions of product removal, each company or plant will make a procedure outlining the steps it will take to effect the efficient removal of product that is in violation of regulatory requirements.

1. Responsibilities: The procedure will identify the participants on a recall or crisis team and the responsibility of each participant in the event of product recovery action. This typically is a cross-functional team with a representative from Management, Quality Assurance, Production, Sales or Customer Service, Warehousing, and Distribution as well as internal or third-party legal counsel.

Author's Note: It is a good idea to maintain a file or record of each team member with 24-hour, 7-day contact information such as business phone number, business e-mail, cell phone, home phone, and e-mail address [7]. An example is provided in Table 2.2.

---

2. Duties of the team participants may include the following:
   a. A Recall Coordinator who will assemble the team, gather information about the situation, identify the type of recovery (i.e., Market Withdrawal, Recall, etc.) and classification (Class I, Class II, etc.) of action involved, provide the team with factual progress reports during the action period, maintain a log of all events and when they occurred as well as the company response, act as liaison between the company, suppliers, and USDA or FDA, participate in all discussions with regulators with input from executive management and legal, coordinate the utilization of outside resources such as forensic labs, micro labs, or consultants as needed,

Table 2.2   Recall Team Contact List

Cleo's Foods
Recall Team Contact List

Name	Title	Office Phone	Office e-mail	Office Cell	Home Phone	Personal Cell
Cleo Katt	President	(800)328-3663	cleo@cleos.com	(800)555-8677	(800)668-7137	(800)467-5337
Michael John	QA Manager	(800)328-3663	mike@cleos.com	(800)555-8677	(800)668-7137	(800)467-5337
Buckey Katt	Sanitation	(800)328-3663	buck@cleos.com	(800)555-8677	(800)668-7137	(800)467-5337
Elliot Katt	Production	(800)328-3663	elliot@cleos.com	(800)555-8677	(800)668-7137	(800)467-5337

Note: The same type of format and information can be used for customer emergency contact information.

coordinate all Practice Product Recoveries, and provide a summary of product recovered and disposed.

b. Monitor customer or consumer feedback from the Customer Service, and assess the nature of consumer questions/comments. Communicate with the Recall Coordinator if communication from customers, consumers, suppliers, or regulatory agents indicates a potential need for product recovery action. Maintain a list of all brokers, distributors, and customers and act as primary customer/distributor contact in the event of a recovery or recall. The list will include primary and secondary contacts; phone numbers, fax numbers, address, and emergency/weekend numbers.

c. Provide updates from suppliers to ensure effective communication regarding ingredients in nonconformance.

d. Make the final decision to proceed with the recall, and communicate the decision to company legal counsel. This is usually done by the plant owner or company executive.

e. Ensure that manufacture of nonconforming product is discontinued until corrective action brings the product or process back into control.

f. Control finished product through the transportation chain. If product is in transit, stop delivery or arrange for product to be held at the first possible drop-off point. Arrange for return of recalled product, on QA Hold, keep adequate records on returns by identification code (i.e., code date) and quantity, and coordinate the return of product to the appropriate assembly plant or distribution center.

g. Contact with customers and consumers to notify them of the actions being taken by the company.

3. Action Plan: If information available from supplier, customer, or consumer information clearly indicates a potential food safety risk, the Recall Coordinator should immediately initiate the recall process. Place all remaining or suspect component(s) and finished product containing suspect component(s) on internal hold.

a. If ingredients are the source of the problem, work with the suspect component supplier to determine the need to contact the FDA or USDA to identify a recommended action plan.

b. Prepare and distribute a product trace and removal order in accordance with the Recall Team's decisions.

Author's Note: One of the primary factors in conducting an effective recall is having an effective code dating system and good records to trace ingredients and product [9]. Having complete documentation of ingredient lots into finished product, clear and legible code dates on finished product, detailed records of quantities produced, and ability to verify the location of product through distribution may make the recall more efficient and can narrow the scope of the recall [5].

c. The Recall Coordinator, as previously designated, notifies the FDA or USDA Emergency Response Team, if appropriate and as advised by legal and executive management, by phone about the recall decision. Notification will occur within 24 hours of determination of action type and class. The notification should include: identity of the product, reason for the removal or correction, and the date and circumstances under which the product deficiency or possible deficiency was discovered; evaluation of the risk; total amount of suspected product produced and/or time span of the production; total amount of suspected product estimated to be in distribution channels; copies of actual or proposed communications; proposed recall strategy; and name and number of responsible firm official. Forward updated recall status reports to the FDA or USDA, as appropriate, once the appropriate plant or corporate and legal counsel has authorized them.

d. The Recall Team will work with the appropriate internal staff to gather manufacturing and shipment records. These records will include at a minimum the following information: product quantity produced, product quantity shipped to outside warehouse or customer, product code date numbers, and other reports as requested.

Author's Note: When preparing this information, consider any work in process product (WIP), returns, or rework used that may be suspect.

e. Communicate necessary information to outside warehouse or distribution centers furnishing recalled product and pertinent return information.

f. Internal communications are important in the event a Recall situation as it is likely to be covered by the news media. Internal communications to company employees through confidential bulletins or secure e-mail will be handled by the Recall Coordinator to present factual information about the event.

Author's Note: Cellular phone calls are not secure and can be electronically intercepted. All phone conversation regarding product action should be made on secure land-based lines.

g. Notification to Customers or Consumers should be done through Customer Service and/or Sales to inform them of the decision to recall product. This will begin with phone contact and may be followed up with a letter confirming the appropriate product information. Arrange to have form letters prepared, approved, printed, and delivered, as instructed by the Recall Coordinator to consumers.

Author's Note: Just as it is important to have a record of internal team member contact information, it is also a good idea to have emergency contact information for customers or customer representatives as shown in Table 2.2.

---

h. In the event the media calls for information, it may be best to have a call sheet for the plant receptionist to take information until the appropriate person can respond. An example is presented in Table 2.3. Prepare appropriate media notification (radio, newspaper) as well as statements for news media calling for information. It is a good policy to have a prepared release in the event of a recall, as identified in Table 2.4 [4]. The main purpose is to avoid the circulation of erroneous information. Legal counsel should review and approve a news release or a position statement on the recall and be ready to issue it when such situations arise.

4. A Recall effectiveness check is conducted to establish the progress made during the process to verify that all means are used to identify the location of the suspect product. Based on regulatory agency directions, the following information will be developed, depending on the recovery action being taken [2]:

a. Class I Recall: In this situation, a Level A effectiveness check is warranted. Level A requires that 100% of known direct accounts and subaccounts and, if necessary, the consumers that are to be contacted.

b. Class II Recall: This situation warrants a Level A, B, or C effectiveness check. The Recall Coordinator will make the decision as to which effectiveness check level is employed at the time of the recall, depending on the nature of the problem that caused the recall and the circumstances surrounding it.

   i.   Level A has been defined above.

   ii. Level B involves any percentage of direct accounts or subaccounts to be contacted that is greater than 10% and less than 100%.

   iii. Level C requires 10% of the total number of direct accounts, and two subaccounts of each direct account to be contacted.

Table 2.3   Media Call Worksheet

Cleo's Foods *MEDIA CALL WORKSHEET*	
RECEIVER NAME:	
CALLER NAME:	
ORGANIZATION:	
E-MAIL ADDRESS:	
DIRECT DIAL:	
CALL DATE:	CALL TIME:
NATURE OF INQUIRY:	
RESPONSE:	
FOLLOW-UP:	
OTHER:	

   c. Class III Recall: This situation warrants Level C, D, or E effectiveness checks. Level C has been defined above.

      i. Level D requires 2% or less of the total number of direct accounts or one per field office in whose area direct accounts are located, whichever is greater, and one subaccount for each direct account to be contacted in both cases.

      ii. Level E requires no effectiveness checks.

   d. Market Withdrawal: This situation warrants Level C, D, or E effectiveness checks as defined above.

   e. Stock Recovery: This situation, by definition, is Level E effectiveness check since the whereabouts of the recalled product are known. No effectiveness checks will be done when this level is designated.

Table 2.4 Media Statement Example

Cleo's Foods *Media Statement*
Cleo's Foods number one priority is food safety, and our food safety record is exemplary. We are also greatly committed to providing the highest-quality product to our customers. Our emphasis is on preventing pathogens from entering our system and we devote an extraordinary amount of resources—both financial and human—to ensure safety and wholesomeness. However, we operate in an environment where pathogens are ubiquitous in the environment. Recently, we were informed by U.S. Department of Agriculture officials of laboratory tests that indicate the discovery of **[name of pathogen]** in one of our products. We acted immediately, through our Food Safety Task Force, to identify the source of the bacteria and prevent its recurrence. Cleo's Foods has notified all of our customers who may have received **[name of product; production code]** that we are voluntarily withdrawing this particular production lot from wholesale and retail channels. We are increasing the frequency of our microbiological testing, and once we have determined the source of entry and verified, through further laboratory testing, that our product is free of **[name of pathogen]**, then we will resume production and distribution. Cleo's Foods remains dedicated to food safety and will continue to be a leader in development and implementation of food safety technology throughout our system.
Contact: Cleo Katt, President

5. The Product Recall should be terminated when it has been determined that all reasonable efforts have been made to remove or correct the recalled product in accordance with the recall strategy, and when it is reasonable to assume that the recalled product has been removed and proper disposition or correction has been made commensurate with the degree of hazard warranted by the recall classification and individual circumstances. If the FDA or USDA has participated in the product recall, the authority to terminate the recall rests with the recalling firm as recommended to the agency. Upon termination of a recall, a formal report will be prepared and presented to the appropriate regulatory agency. The report will be prepared by the Recall Coordinator and will include item name and number, date recall initiated, total product produced, total product recovered, disposition of recovery, and date of recall termination. The report will be reviewed for input by legal counsel. A debriefing with the Recall Team and company management will be conducted to review the effectiveness and needed corrective actions.

Practice Recovery

In an effort to ensure that all participants know their assignment and can carry it out efficiently, it is always a good idea to practice the procedures identified in the recall plan. Many plants conduct a "Mock Recall" to test the system. This allows the company to verify that people know what to do in the event of a recall and to identify areas in the system where there are flaws or gaps that would not result in thorough or timely removal of product.

Author's Note: Some companies prefer not to use the term "Mock Recall" due to the obvious negative connotation of the word *recall*. An alternative term that will be used in the sections that follow is "Practice Product Recovery."

Plants should test their system at least once per year and more often if the practice product recovery results are less than 100% of products identified within two hours of initiation. There are two types of practice product recoveries that should be conducted, one for finished product and one that begins with the identification of a raw material that is traced to finished products. The latter is more challenging and requires that the plant maintain very thorough records on the use of raw material lots.

When a practice product recovery is conducted, the plant Recall Team will convene to review the information on the product or raw material to be traced and begin the records review and information gathering. If the practice product recovery involves only product, determine the quantity produced and shipment locations. The plant team will consider whether more than one production line, rework, mixed code repack, or work in process (WIP) is involved.

If the practice product recovery involves an ingredient, determine the following: How much was received? When was the ingredient used? What products was it used in? How much was used? How much remains? When the above are determined, ensure that the quantities of ingredient balance to current. The team will select *one* product from one code date that the ingredient was used in, and that will be the product for the continued practice product recovery. The plant team will determine quantities shipped and shipment locations. If product goes to an internal or third-party distribution center, the plant with product quantity and code date information should contact them. When involved, the distribution center will pull together all records to determine quantities received and locations shipped. They will notify the plant so that quantities produced and quantities received by the centers can be reconciled.

Author's Note: Unless previously arranged, do not contact brokers, distributors, or customers during practice product recoveries. The process terminates with company-controlled or third-party distribution centers.

When all information is received from all participating parties, and the information is compiled on the Product Recovery Worksheet (Table 2.5), the practice product recovery will be terminated, the percentage of product will be calculated, and the time of the recovery determined. This information will then be reviewed within 24 hours, with all participants to evaluate effectiveness toward the goal. The practice product recovery effectiveness goal is 100% of quantity produced. The completion time effectiveness goal is two hours from the recovery start time. The ingredient recovery effectiveness goal is ±5% of the received ingredient weight. Recovery rates that do not conform to the established quantity and time goals would result in an additional recovery being conducted to verify corrective actions.

It is hoped that incorporation of all of the food safety best practices that follow in this book will help companies avoid having to implement a product recall. However, every company will be wise to have a well-rehearsed plan in the event that it is needed.

Table 2.5 Product Recovery Worksheet

Cleo's Foods *Product Recovery Worksheet*								
Please Identify One								
Recall		*Practice Product* *Recovery*			*Market* *Withdrawal*		*Stock Recovery*	
Recall Number:					Date:			
Product:		Product Number:			Date Produced:		Code Date:	
Person Requesting Recall/Withdrawal:					Person Requesting Recall/ Withdrawal:			
Reason for the Recall/Withdrawal:								
Total Number of Cases *Under Recall/* *Withdrawal:*			*Total Number of Cases* *Distributed at Time of* *Recall:*			*Total Number of Cases* *Remaining in* *Companies Possession:*		
List All Codes and Number of Cases for Each Code Being Recalled/Withdrawn								
Product *Code*	*Code* *Date*	*# of* *Cases* *per* *Code*	*Product* *Code*	*Code* *Date*	*# of* *Cases* *per* *Code*	*Product* *Code*	*Code* *Date*	*# of* *Cases* *per* *Code*
Comments:								
Location of All Cases and/or disposition: (Attach copies of shipping documents) FOR DETAILED LIST OF RETURNED PRODUCT SEE PRODUCT RECALL INVENTORY REPORT								

Continued

Table 2.5 (*Continued*)   Product Recovery Worksheet

Cleo's Foods Product Recovery Worksheet		
Product Code	*Number of Cases*	*Location and Disposition*
Total Number of Cases Recovered:	Percent Recovered:	Date and Time Recall/ Withdrawal Completed:
Completed by:		
Confidential Commercial Information		

References

1. Gould, Dr. Wilbur A., *CGMP's/Food Plant Sanitation*, 2nd edition, CTI Publications, Baltimore, 1994, chaps. 2 and 3.
2. Stauffer, John E., *Quality Assurance of Food Ingredients Processing and Distribution*, 3rd edition, Food and Nutrition Press, Trumbull, 1994, chap. 6.
3. Troller, John A., *Sanitation in Food Processing*, Academic Press, New York, 1983, pp. 391–396.
4. Anon, *The Key to Managing a Business Crisis*, National Meat Association and Edelman Public Relations, 2001, p. 4.2.
5. American Meat Institute Foundation, *Meat and Poultry Recall: Are the Rules Different under HACCP*, presented at Washington, D.C., September 18–19, 1997.
6. American Meat Institute Foundation, *Recordkeeping and Recall Seminar*, presented at Charlotte, N.C., February 26–27, 2001.
7. *Guidelines for Product Recall*, Grocery Manufacturers of America, Washington, D.C., 1983.
8. U.S. Department of Agriculture, FSIS Directive 8080.1, *Recall of Meat and Poultry Products*, Washington, D.C.
9. Tybor, Philip T., William C. Hurst, A. E. Reynolds, and G. A. Schuyler, *Quality Control: A Model for the Food Industry*, University of Georgia College of Agricultural and Environmental Sciences Cooperative Extension Services, 2003, p. 14.
10. U.S. Department of Agriculture, *Rules of Practice*, Washington, D.C.

Chapter 3

Microorganisms of Food Manufacturing Concern

If it don't stink, stuff it!

—**Anonymous food plant employee**

They go by many names: germs, bacteria, microorganisms, microbes, even "bugs." They have been the main focus of several Hollywood movies, including *The Andromeda Strain*, *Outbreak*, and *The War of the Worlds*, where they were responsible for saving the human race. They have been on Earth for millions of years, longer than man.

Microorganisms

Microorganisms are biological entities, and they can be a benefit or a potential hazard to humans and to the food manufacturing industry. Bacteria represent the largest group of microorganisms [16]. Most bacteria are harmless; in fact, some provide benefit to humans by protecting the skin and nasal passages and also aid in the digestion process. Some also benefit the food industry when they are used for production of cultured items such cheese, yogurt, and fermented sausage. However, they can pose a threat to humans and to the industry when they result in foodborne illness and food spoilage. Microbiological contamination of food may

result in product spoilage, reduction in shelf life, or foodborne illness. It is important to understand food-related microorganisms, with regard to growth needs
and environmental requirements to better understand their control mechanisms.
Fortunately, the food industry has evolved from the dinosaurs who truly believed
that if something does not stink, it is OK to stuff to a more science-based understanding of microorganisms that impact the food industry. The common microorganisms that will be reviewed in this chapter, as they relate to foods, are indicator
organisms, spoilage organisms, and pathogens. The focus will be on understanding
their functional needs as well as control, particularly as it relates to sanitation.

What are bacteria? As stated before they are living organisms of various shapes
and sizes that have the same basic needs as more complex organisms. They all
require food, moisture, and time for growth. However, different organisms can
require a range of temperatures and oxygen requirements for growth, a topic that
will be reviewed in this chapter. The growth of organisms involves several phases.
The first is the lag phase, in which the organism produces enzymes to reproduce,
provided environmental conditions are suitable. If the conditions are not suitable,
no enzymes are produced. There is no change in the bacterial numbers at this
point [19]. Bacterial growth requirements will be covered in greater detail in the
next sections. During the second phase, or the log phase, bacterial reproduction
begins, and the time between cell division is called the generation time. This time
depends on the environmental conditions [19]. When bacteria grow, they do not
grow arithmetically; that is, they do not grow from one cell to two cells to three
cells. Rather, they grow geometrically, or by binary fission, where by each cell
divides so that one cell becomes two, two cells become four, four cells become
eight, and so on, until a colony of millions of cells can form [7]. Substantial growth
of an organism can occur on a small, even microscopic, mass of food. Under ideal
conditions, cell division takes approximately 15–20 minutes, so if a food contact
surface has food material on it and becomes contaminated with bacteria, it will
not take long in a production shift for the surface to bear millions of cells. This is
illustrated in Figure 3.1.

The final phase is the death phase, in which there is a decline in the overall bacterial population. As bacteria grow and consume nutrients present in the environment, their metabolism produces waste products. Over time, these waste products
become toxic, resulting in cell intoxication and death [2].

Bacterial Requirements

Bacteria can be found almost everywhere in nature, and they come in various sizes
and forms. They have varying growth requirements as do more complex living
organisms. They all have basic needs of food, moisture, and time for growth; however, some have specific needs that will also be discussed.

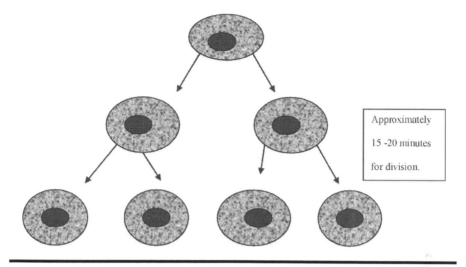

Figure 3.1   Illustration of binary fission.

Moisture

Water activity (Aw) is very important to bacterial growth, as all bacteria require moisture. Food products contain varying amounts of water, and knowing the water activity of the product will help provide an indication of the potential for microbiological growth. It is important to note that water activity is not same as the moisture percentage in a product. The percent moisture is the total amount of water in the product, whereas water activity is the amount of free water available for bacterial growth. It is measured as the vapor pressure of solute over the vapor pressure of pure water. If pure water is expressed as a value of 1.0, then water activity is measured as a percentage of pure water and it is generally less than 1.0 (e.g., 0.93). A level of 0.95 or higher means that water is available for growth. The order of water needs is bacteria (highest), yeast, and mold (lowest) [4,15].

Temperature

The ideal temperature growth range for most microorganisms is 40°F to 140°F; however, as previously mentioned, different bacteria have different growth temperature requirements. These are identified in Table 3.1.

Psychrotrophs and mesophiles are the greatest concern to the food industry, and they are the growth ranges for spoilage and pathogenic organisms. Psychrotrophs grow well at refrigerated temperatures and best at room temperature. These organisms can survive freezing but fail to grow above 90°F. Mesophiles are moderately heat loving and grow at room temperature but grow best as they approach body temperature. Mesophilic growth is slowed below 40°F and above 140° F. Thermophiles are heat-loving organisms. These are important factors in

Table 3.1   Bacterial Temperature Ranges

	Temperature Growth Range	Ideal Growth Temperature	Organism Examples
Psychrophiles	−5°C–20°C 25°F–60° F	15°C 60°F	Vibrio
Psychrotrophs	0°C–35°C 32°F–95°F	24°C 75°	Listeria monocytogenes Pseudomonas, fungi
Mesophiles	15°C–48°C 59°F–118°F	37°C 98°F	Pathogens Salmonella Campylobactrer
Thermophiles	40°C–70°C 104°F–158°F	55°C 130°F	Fungi Pathogens Bacillus cereus, Clostridium

Source: Frank, Hanns K., *Dictionary of Food Microbiology,* Technomic Publishing Company, Lancaster, 1992, pp.

maintaining foods at temperatures that will least likely accelerate the growth of microorganisms, whether they are being stabilized after the cooking process or being held at specific temperatures in cooking vessels.

Oxygen

Here again, different organisms have different oxygen requirements. Those that require and grow best in an oxygen environment are referred to as aerobic bacteria. This group is represented by many of the spoilage and pathogenic organisms. This is why some food items are vacuum-packaged, where the air is removed from the package, or the oxygen is replaced by a mixture of carbon dioxide (CO_2) and nitrogen (N_2) gas in a mixture that inhibits the growth of anaerobic bacteria. Food manufacturers who use this technology have to be aware that by eliminating the growth of aerobic bacteria they may be selecting for another group of bacteria referred to as anaerobes. These bacteria survive in an atmosphere where there is little or no oxygen. Unfortunately, they contain some of the most severe pathogens such as *Clostridium perfringens* and *Clostridium botulinum*. A third category is a group called facultative anaerobes. These are organisms that can adapt to conditions that may or may not include oxygen.

Differentiation

There are several means of differentiation between organisms, including shape, Gram staining, spore formation, and metabolism of nutrients. This information can be helpful when identifying the organisms present in a food processing operation and when designing control and elimination strategies to protect the safety and quality of the food.

Bacteria Shapes

Bacteria appear as one of five basic cell shapes when viewed under a microscope [16]. They are cocci or round, bacilli or rod shape, spirilli, vibrio or "comma" shaped, and filamentous.

Gram Staining

One of the means of differentiation among organisms is through the process of Gram staining. Danish physician Christian Gram developed the procedure of Gram staining in 1884. The process is used to differentiate types of bacteria. Gram-positive bacteria retain the crystal-violet stain, and Gram-negative bacteria lose the crystal-violet stain, resulting in a red/pink appearance. It is important to know this when considering the use of sanitizers as some sanitizers are more effective against Gram-negative bacteria while others are more effective against Gram-positive organisms. Chapter 7 will cover the application of sanitizers in greater detail.

Spore Formation

Resting bacterial cells are referred to as vegetative cells, which have an active metabolism and may or may not be growing [15,19]. Some vegetative cells are non-spore formers, mostly round or coccid organisms and many rods. Others, including some rods, can form spores. Spores are considered the resting or dormant phase of the cell and are similar to plant seeds [15]. Spores can survive a wide range of conditions, including heat, cold, and chemicals, and can subsequently serve as a contaminant that can cause problems if conditions become optimal for growth.

Nutrient Metabolism

Organisms can also be categorized by the nutrients they metabolize. This is demonstrated in Table 3.2 and can be useful, depending on the product manufactured.

Nonpathogenic Microorganisms

Another means of classifying bacteria is their ability to impact humans and disease. The group of bacteria that causes disease in humans is referred to as "pathogens," and they will be covered later in this chapter. Bacteria that do not cause disease are referred to as "nonpathogenic bacteria," and may also be called "indicator organisms" as they are used to indicate the levels of cleanliness. Generic bacteria are

Table 3.2 Nutrient Metabolism

Metabolic Category	Nutrient
Proteolytic	Break down proteins
Lipolytic	Spoil lipid-containing foods
Saccharolytic	Break down pectins
Amylitic	Break down starch to sugars
Cellulytic	Break down cellulose to simple carbohydrate compounds

found almost everywhere in the environment. Many are vital to the equilibrium of nature. Some aid in digestion, protect our skin or nasal passages, and are used in the manufacture of foods such as cheese, yogurt, and dry sausage. For example, yeast is used in bread manufacturing to help the dough rise. *Lactobacillus delbrueckii* subspecies *Bulgericus* is used in the production of yogurt, and their growth in the milk medium is under a controlled environment. Even wine relies on bacteria; the unique and complex flavor characteristic of Sauterne is the result of *Botrytis cinerea* (also known as Noble Rot) that forms on the vine root. Some do not result in food spoilage but are indicators of conditions that could support the growth of spoilage or pathogenic organisms. These are referred to as indicator organisms, and they include coliforms and generic *E. coli*. These are Gram-negative rod-shaped enteric bacteria that can be found almost everywhere in the environment. They are typically aerobic, but can be facultative. They may also be considered as quality drivers; that is, the initial numbers of bacteria, types of bacteria present, and the storage conditions of the finished product may have a bearing on the performance of the product over its shelf life. While they may not result in foodborne illness, large numbers of coliforms are not desirable in foods and may indicate that the production environment was not sanitary. The majority of *E. coli* organisms are harmless enteric bacteria from human or animal sources [18]. Though they may not cause disease, they are used as indicators of possible enteric or fecal contamination. Generic *E. coli* is used as a fecal indicator in water due to its survival period [17].

Spoilage Organisms

Contamination of food products with many microorganisms does not result in foodborne illness; however, it may be the underlying cause for the spoilage of food products and thereby result in products that are undesirable to eat. While this group of organisms is easily eliminated through cooking, they are responsible for spoilage of product in the refrigerated state. This group can include the following.

Yeasts are unicellular organisms that also fall in to the category of fungi (but not bacteria). They are round to oval in shape, nonmotile, have no chlorophyll, and are widely distributed in nature [16]. They reproduce by budding or spore formation. Most are not pathogenic, but because they live on sugar and starch, can result in spoilage of foods through a fermentative process. They produce CO_2 and alcohol, which aids in production of bread dough, beer, and vinegar, but spoils jelly, honey, and syrup [16].

Mold is a multicellular organism also referred to as fungi (not bacteria). They are nonmotile and filamentous [16]. It can be useful in the production of soy sauce and some cheeses. It grows best at room temperatures and can grow under refrigeration. Mold growth results in filaments or hyphae, which, as they mass, become the visible presence of mold, or mycelium, which spreads across food. Spores from mold travel on air and can form new colonies where they land [15]. Mold is generally not harmful unless they form mycotoxins. Mycotoxins are secondary metabolites of fungi and are the result of the natural metabolic outcome of typical food mold genera *Aspergillus*, *Penicillium*, and *Fusarium* [24]. They are resistant to heat and can be harmful to animals. Some mycotoxins such as aflatoxin, which is a by-product of *Aspergillus flavus* mold found on peanuts, may be carcinogenic to humans [6]. Yeast and mold are not typically factors in meat spoilage but can be factors in other foods [19].

Pseudomonads are a common food spoilage organism, especially on the surface of ground meat. They are Gram-negative obligate aerobic rods and psychrotrophic. They are water loving. They are not associated with foodborne illness but are responsible for slime formation, off-odor, off-color, and greening in meat. They can cause cellular greening. There are two specific subspecies of particular food manufacturing concern. *Pseudomonas fluorescens* is found in water, soil, and animal intestinal tract. They can result in off-flavor and odor, green rot, or bluish cast. *Pseudomonas fragi* are found on the surface of meat, in milk, and on eggshells. They can result in off-color and off-odor.

Pathogens

Pathogenic bacteria make up a class of microorganisms that have been associated with human foodborne illness. They can result in consumer sickness, hospitalization, fatality, recall, liability, and loss of business. For this reason, they are of great concern to the consuming public, the food industry, and regulatory agencies. Specific pathogenic organisms that have been most often associated with illness and thus labeled as organisms of public health concern are *Listeria monocytogenes*, *Salmonella*, *E. coli* O157:H7, *Staphylococcus aureus*, *Campylobacter*, and *Clostridium* organisms. Others of concern include *Bacillus cereus*, *Yersinia enterocolitica*, viruses, and parasites. All will be detailed in this section of the chapter.

Listeria monocytogenes

Listeria is a Gram-positive rod-shaped bacillus, non-spore-forming and motile. It is facultative, meaning that it can live in an aerobic (with oxygen) environment or adapt to an anaerobic (without oxygen) environment. It is often referred to as a ubiquitous organism, meaning it can be found frequently throughout the environment, especially in soil. *Listeria* is also commonly found in human and animal feces, raw vegetables, and cheeses made from raw milk. There is a high incidence rate in raw meat and poultry. The pathogenic form, *L. monocytogenes*, is readily found in raw meats. It is more heat resistant than other pathogens but is destroyed by cooking. Because of the ubiquitous nature of *Listeria* in the environment, it is particularly challenging to control in the food manufacturing operation because it can grow in a wide range of conditions. The organism is slow growing and grows well in a temperature range from 32°F to 113°F and survives freezing, although it will stop growing below 31°F. It does not compete well with other cold-tolerant organisms [25]. *Listeria* can grow in a pH range from 5.2 to 9.6, is salt tolerant between 5% and 10% and has a water activity (Aw) > 0.93. Because of the significance of *L. monocytogenes* in the food industry, it has been given its own chapter on control. Control of *L. monocytogenes* is covered in greater detail in Chapter 4.

Salmonella

Salmonella is a genus of the family Enterobacteriaceae and is a Gram-negative, facultative aerobic rod with flagellate motility. It too is an intestinal bacteria often found in the fecal material of birds, livestock, and pets, but it is also ubiquitous in dirt. There are more than 2200 serotypes, and all are known to cause disease in humans; however, the two most common are *S. typhi* and *S. paratyphi*. It survives at pH 4–8 and an Aw > 0.94. The temperature growth range is 41°F–115°F [19], and it will survive freezing. It is generally not an environmental contaminant like *Listeria*; however, control involves separation between raw and cooked. The disease is called salmonellosis, and it can affect all humans regardless of age, sex, or health status. The symptoms include headache, nausea, abdominal pain, vomiting, and non-bloody diarrhea.

Escherichia coli O157:H7

This is a strain of *Escherichia coli* bacteria most often found in the intestinal tract of cattle. It is a Gram-negative rod and a facultative aerobe. It grows optimally at 50°F–108°F, pH above 4.5, and Aw < 0.92 [19]. Human illness is caused by the verotoxins produced by the organism, and the result can be as serious as bloody diarrhea or as severe as a condition called hemolytic uremic syndrome, or HUS. HUS is characterized by failure of the kidneys and often results in the need for dialysis. It is believed that as few as 10 organisms can cause illness in humans through colonization of the intestinal tract. Transmission of *E. coli* O157:H7 is through three distinct sources: undercooked hamburger, cross-contamination from raw beef, or vegetables fertilized with unprocessed cow feces. It is an adulterant in raw ground beef. Ground beef needs to be fully cooked (to 160°F). There have been some associations with fecal contamination from wild animals such as deer or elk.

Staphylococcus aureus

Staphylococcus aureus is a facultative non-spore-forming aerobe. It is coccid in shape, and appears as grape-like clusters when magnified. The organism is killed by cooking but produces heat stable enterotoxins, which cause a large number of foodborne illnesses. The toxin is destroyed during heat treatment in canning operations. *Staphylococcus* grows at temperatures between 44°F and 115°F, pH > 5.2, and has low moisture needs (Aw of 0.86) for growth [19]. Overall, *Staphylococcus* is a poor competitor but can be hazardous if introduced at a post-lethality stage where there are few competitors. Sources of this organism are the human nose or throat discharge, infected cuts or wounds, burns or boils, and pimples. Hands are a common mode of transmission, and hence the importance of hand washing. It is present in low numbers in raw meat. Presence is not as important as the numbers of organisms from growth. It requires high numbers (10^6) to produce toxin [19], and this is often an indication that the material has been time and temperature abused [11]. Testing of raw incoming meat for *Staphylococcus* can be used as a measure of supplier GMPs. Foodborne illness caused by this organism is often associated in high-protein foods that have been "rewarmed."

Campylobacter

Campylobacter is a Gram-negative, spiral-shaped, microaerophilic (low oxygen requirements) organism that grows best around 86°F–117°F. The minimum pH for growth is 4.9, and the minimum Aw is 0.98. It is an enteric pathogen of warm-blooded animals but is an organism of concern in raw poultry. Cooking kills it, and illness is usually as the result of undercooking or cross-contamination. It is suspected to be one of the most common causes of foodborne illnesses; fortunately, infections in humans are rarely fatal [6].

Bacillus cereus

Bacillus cereus is a Gram-positive spore-forming rod. This organism is a facultative obligate aerobe, grows ubiquitously in the environment, and has low acid tolerance. While *B. cereus* is associated with a low incidence rate of foodborne illness, it is generally associated with foods held in warming trays, especially rice, and it is controlled through proper cooking and chilling [6].

Yersinia enterocolitica

Yersinia enterocolitica is a Gram-negative bacillus associated with water and foods. It is a hardy organism that survives alkaline pH, is tolerant to salt, and withstands freezing. It is not a psychrophile but can grow in temperatures as low as 40°F. Heating to a sufficient temperature for lethality and rapid stabilization through chilling will control growth.

Clostridium

The two organisms associated with foods are *Clostridium botulinum* and *Clostridium perfringens*. *Clostridium botulinum* is a Gram-positive straight to curved motile rod. It grows best between 38°F and 115°F at a pH above 4.7 and an Aw above 0.94 [19]. It is a strict anaerobe and is ubiquitous in nature, especially in soil. It produces a heat-resistant spore that produces a toxin that has the potential to produce eight different toxins that impact neurological processes in humans through botulism intoxication. This is typically not an organism that is associated with sanitation; rather, it is a function of under-processing or a consequence of a failure to follow cooling stabilization in cooked products, especially in home canning operations. Fortunately, *C. botulinum* intoxication is rare; however, the mortality is high, and therefore it must be considered a severe hazard.

Clostridium perfringens are Gram-positive anaerobic spore-forming rods. Growth is between 59°F and 122°F, above pH 5.5, and Aw 0.95 [19]. They are one of the leading causes of foodborne illness and can grow to large numbers of vegetative cells under proper growth conditions that support germination of the spore state. This organism is often found in raw food products and survives cooking; however, temperature abuse increases growth. Further, it is controlled by proper post-lethality chilling (stabilization).

Viruses

Viruses are obligate intracellular parasites. They require a living host so that they can invade living cells and begin replication, and as such they do not grow well in foods. However, viruses such as Hepatitis A and Norwalk are concerns within the food industry, specifically the food service industry. Viral spread is usually through the fecal-oral route, which means that they can be spread as result of poor GMPs (not washing hands after using the restroom) or through sneezing and coughing by infected persons. They can also be transmitted through contaminated seafood and shellfish. Hepatitis causes liver infection and results in nausea, fever, and jaundice; fortunately, it is rarely fatal. Norwalk is the most common foodborne virus and is spread by contaminated water or by person-to-person contact. The most effective means of control are GMPs (covered in Chapter 7), hand washing, and not allowing infected individuals to work around food [6].

Bacterial viruses are known as bacteriophages, and they are widely distributed in nature. They are composed of DNA and RNA as well as several proteins. They do not reproduce by binary fission; rather, they attach themselves to host organisms and deposit their nucleic acid. Many phages then form inside the host and are subsequently released. These continue the process by inoculating more cells. These can be a particular problem for cultured dairy products as they can result in failure of the starter culture. They are controlled through sanitation and GMPs [15].

Parasites

Parasites are not generally related to improper food plant sanitation and are rarely a problem in the United States. The most familiar parasite is *Trichinella spiralis*, or Trichina, a roundworm found in pork. If undercooked pork is consumed, Trichina infects the muscle tissue of the host. However, common freezing practices render raw pork Trichina-free, and the parasite is killed by cooking. Other parasites include amoeba (water), Toxoplasma (cat feces), and Giardia (water); however, they are rarely associated with food consumption in the United States.

Foodborne Illness

Foodborne pathogenic organisms cause illnesses in two different ways. The first is through bacterial intoxication, ingestion of food containing a toxin formed by bacteria. In the cases of illness associated with *Staphylococcus aureus* or *C. botulinum*, it is the toxin that causes the illness. The second is through bacterial infection, where the illness occurs from ingesting live bacteria that grow and cause illness [3]. The number of cells required to cause illness is referred to as the infectious dose and the number required may vary depending on the specific organism. *L. monocytogenes* has a relative low infectious dose in susceptible individuals, *Salmonella* has a variable infectious dose, and *E. coli* O157:H7 has a low infectious dose [19].

One of the most frustrating calls that can come into a food plant or company is that of a consumer calling to report an illness. Though many individuals believe foodborne illness is caused by the last thing eaten or always the result of eating meat products, this is not necessarily the case. Illness may be caused by a variety of foods that are not of animal origin. It is always a good idea to treat each call seriously and have a call line or customer service to log illness calls. Since a foodborne illness outbreak is described as an incident in which two or more people experience a similar illness after the ingestion of a common food, the logged illness calls should be tracked for follow-up. It is also helpful for the person taking the call to understand the symptoms to aid in identification. Table 3.3 illustrates the illness caused by pathogens, the onset time, and symptoms associated with the disease.

This information is not intended to be used for the purpose of diagnosis; only a doctor should do this. It is intended to be used as a means of classifying information coming from customer calls.

Microbiological Control Methods

Control of spoilage or pathogenic microorganisms in a food manufacturing operation requires many steps, what is often referred to as a "multiple hurdle" approach.

Table 3.3   Common Foodborne Illness, Onset, and Symptoms

Organism (Illness)	Time to Onset	Symptoms
Listeria monocytogenes (Listeriosis)	7–30 days	Fever, nausea, headache; meningitis-like symptoms, septicemia. *Can be fatal if not treated promptly.*
Salmonella (Salmonellosis)	1–4 days	Nausea, fever, abdominal pain, diarrhea, dehydration.
E. coli O157:H7	2–10 days	Cramps, fever, vomiting, profuse watery diarrhea. The disease can proceed to hemolytic uremic syndrome (HUS), which can result in kidney failure and dialysis.
Staphlococcus aureus	1–6 hours	Nausea, vomiting, diarrhea, cramps.
Campylobacter (Campylobacteriosis)	3–5 days	Fever, nausea, vomiting, abdominal pain, diarrhea.
Bacillus cereus (food poisoning)	1–16 hours	Emetic—nausea, vomiting, sometimes diarrhea. Diarrheal—diarrhea, cramps, sometimes vomiting.
Yersinia enterocolitica (gastroenteritis)	1–3 days	Fever, diarrhea, abdominal pain. Appendicitis-like symptoms.
Clostridium botulinum (botulism)	12–36 hours	Fatigue, weakness, double vision, slurred speech, respiratory distress. *Highly fatal if not treated.*
Clostridium perfringens (food poisoning)	8–22 hours	Diarrhea, cramps, sometimes nausea and vomiting.

Source: From Imholte, Thomas J. and Tammy K. Imholte-Tauscher, *Engineering for Food Safety and Sanitation*, Technical Institute of Food Safety, Medfield, 1999, pp. 4–6; Katsuyama, Allen M., *Principles of Food Processing Sanitation*, Food Processors Institute, 1993, pp. 65, 77.

This means that many strategies will be employed, primarily to prevent microorganisms from having an opportunity to come into the facility, establishing themselves in the facility, or growing if they become established. It also involves elimination of the organism from ingredients and prevention of post-lethality process recontamination. There are formulating measures that can be followed to affect the pH, moisture level, especially water activity, salt level, and use of inhibitory ingredients such as sodium nitrite and potassium lactate and sodium diacetate. In addition, there are packaging strategies such as vacuum packaging, gas back flush, and use of oxygen scavengers to prevent growth of bacteria that may be introduced post-lethality. In addition, there are post-packaging treatments such as hot water, steam, or pressure post-pasteurization that are effective in destroying bacteria present in the food package. These measures are all effective and part of a comprehensive food safety system. To some extent these measures assume that bacteria are present in the post-lethality finished product. This section will focus on some of the measures to be employed by food facilities to enhance sanitation and prevent contamination of food products post lethality and into packaging.

Incoming Ingredients

Ingredients used in food manufacturing can be a source of microbiological contamination and the hazards vary by the ingredient source. As an example, meat and poultry raw materials, or raw materials from other animal sources, have a higher association with both spoilage and pathogenic organisms. Raw vegetables are also a source of microorganisms [5]. Even if vegetables are blanched, they may be a source of microorganisms as the blanching process is only designed to control enzyme activity, not bacteria, and the vegetables may be exposed to post-blanching contamination. Conversely, dry ingredients like salt or low-acid canned ingredients are not a significant source of either spoilage or pathogenic organisms [5].

Testing of finished product for spoilage organisms or pathogens is only one tool in verifying quality or safety; however, it is not an effective means of ensuring quality and safety. The focus of microbiological analysis should be on components, rather than finished product, to ensure ingredient supplier processes are in control and that product contamination is prevented. Using ingredient hazard analysis, a company can set up microbiological risk categories for each of the ingredients that they use in their operation. All ingredients will be identified for level of microbiological sensitivity based on potential hazard, history, composition, ingredient source, and the product/process into which the ingredient is used. Once this is done, they can establish the requirements for supplier testing and Certificate of Analysis (COA) information as well as internal needs to evaluate incoming raw materials. This will include the frequency with which the ingredients are analyzed, the analysis required, and how the results will be used. Ingredients will be tested to verify that they will not contribute to product contamination and to verify that supplier processes provide conforming ingredients. Disposition of nonconforming ingredients or nonconforming suppliers will be identified in the procedure. The following are the recommended steps to developing the procedure:

1. Develop the criteria for ingredient hazard categories and assign ingredients to each category. Quality Assurance might use prior analysis results, supplier data, or scientific literature to assign microbiological hazard categories according to the following risk criteria:
   a. Category 0: No implication of source of spoilage or harmful microorganisms due to source or processing (i.e., salt, commercially sterile canned sauce, starch).
   b. Category 1: Minimal potential source of spoilage or harmful microorganisms, will receive terminal process step or sufficient heat treatment by end user (e.g., IQF eggs, blanched vegetables, spice, breaders, batters, flour, tortilla).
   c. Category 2: Sensitive ingredients based on historical evidence to suggest the presence of spoilage and pathogenic organisms, but they will be sufficiently cooked during processing to eliminate microbiological hazard (e.g., raw meat and poultry, raw bacon, raw eggs).
   d. Category 3: Ingredients with moderate spoilage or harmful microorganism potential not minimized by process (e.g., natural cheeses, unblanched vegetables, minimally processed fruits).
   e. Category 4: Sensitive ingredients used in Ready-To-Heat/Eat products but may not receive a lethal heat treatment or have a history of association with foodborne illness. Young children, the aged, or infirm who may be susceptible to low pathogen doses might consume these products (e.g., cooked meat and poultry, dried/fermented sausage, cooked bacon).
2. Establish sampling criteria for each hazard category based on the hazard and how the ingredient is used in the finished product. As an example of how this might be applied, natural cheese might fall into Category 1 if it is used in a product that receives a lethality step such as a pocket-type sandwich or a burrito. However, the same cheese might fall into Category 3 if it is used in a product that does not receive a lethality step, such as an entrée that relies on the end user to fully heat the product. For the most part, the sampling in this step will be used to measure supplier performance rather than to accept/reject criteria as most of the testing will be conducted at receiving, and the time delay in receiving the results does not make it practical to use the results as a means of rejection. However, the results can be used to provide the supplier feedback on how well they are performing and as criteria for a certification process.

   Examples of categories and COA and plant testing requirements are as follows:
   a. Category 0: No samples required, and no COA required.
   b. Category 1: Sample new ingredient or new supplier with initial, probationary shipments; COA requested with each shipment. Sample annually

as verification of the supplier's continued performance. Collect one sample from each code date in a shipment and composite to one sample. Analyze for aerobic plate count (APC), coliform, generic *E. coli*, yeast, and mold depending on the ingredient and the potential for these organisms.

c. Category 2: Sample initially probationary shipments for supplier (or new establishment from approved supplier) and quarterly as verification; COA requested with each shipment. Sample each date code received and analyze composite. Analyze for APC, coliform, and generic *E. coli* and *S. aureus* (in raw meat to be tempered or held at ambient temperature).

d. Category 3: Sample initial probationary shipments for new supplier and then every 5th shipment if the initial samples are in spec. A COA should be required with each shipment. Collect one sample from each date code in the shipment and composite for analysis. Analyze for APC (except natural cheese products), coliform, generic *E. coli*, yeast, and mold. Again, the organisms selected for analysis will depend on the material and how it is used in the finished product.

e. Category 4: These represent the most sensitive of ingredients so it is recommended that these suppliers pass prequalification requirements and an initial Food Safety and Quality audit by QA or a third-party auditor. Some of the prequalification requirements will be a written GMP program, written SSOPs, well-documented HACCP plan, and an environmental sampling plan for *Listeria*. Sample initial probationary shipments, then sample quarterly to verify continued control. A COA is required with each shipment. If there is no COA or no testing by the supplier, the plant will test every load. Collect one sample from each date code, composite, and analyze. Analyze for APC (except fermented sausage and natural cheese products), coliform, generic *E. coli*, yeast, mold, *Salmonella*, and *Listeria*. *S. aureus* may be run on pepperoni or other fermented product. The supplier COA should reflect analysis for the same organisms as listed in (b) above. If the plant decides that pathogen testing is required for Category 4 ingredients, notify those suppliers that their products will be tested for the indicated pathogens if the supplier will not agree to test for them. If the plant tests for pathogens, ensure that the lot tested was not previously used by the plant in product that has shipped, otherwise a positive pathogen finding will subject that product to recall.

Author's Comment: The question of whether to require suppliers of Category 4 ingredients for pathogens such as *L. monocytogenes* or whether the receiving plant should test is a matter for the company to decide. As a food safety professional with a company that supplies ingredients, the writer would prefer to use science to make that determination. If the supplying company has been audited, has effective GMP, SSOP, and HACCP plans along with a thorough environmental monitoring plan and no indications of facility deficiencies that would result in microbiological contamination, there should be no need to conduct pathogen tests on the finished product. If the ingredient is going into a product that does not receive lethality, the company will have to decide whether testing for pathogens will provide them with an added measure of security; however, they must keep in mind that testing does not reflect the condition of the entire lot. If the product is being produced for a customer or customers who require finished product pathogen testing, the company might want to consider ingredient testing as an additional measure for product safety.

3. Develop a list identifying supplier status (i.e., approved, conditional, rejected) for Category 4 ingredients. Inform purchasing and R&D if the supplier status should change.

   a. Require Purchasing to only buy ingredients only from approved suppliers, utilizing the approved specifications. Request specifications for new ingredients from suppliers along with the COA. (*Note:* COAs should be requested by purchasing. If they are not received from the supplier, this does not reduce the supplier's liability for out-of-specification ingredients. Lack of a COA does not limit the use of the ingredient by the plant.)

   b. Ensure that new suppliers provide a General and Continuing Letter of Guarantee before ingredients are received. Corporate Purchasing maintains the Letter of Guarantee file. It will be made available to plants for purpose of audits or regulatory review. The Letter of Guarantee is assurance from the supplier that the ingredient is not adulterated within the meaning of the Pure Food, Drug and Cosmetic Act. Ingredients will not be received in the plant unless purchasing has a Letter of Guarantee from the supplier. Implement an ingredient-testing protocol. Maintain data history of supplier performance to determine supplier certification status. All ingredient lots tested for pathogens are to be placed on QA Hold and not used in production until testing is complete. Procedures must be in place to ensure that ingredient lots previously used in production are not pathogen tested. The QA Manager will evaluate plant analysis results and compare them with the COA received. Results that are out of specification for pathogens in Category 3 and 4 ingredients that do not receive a lethal treatment will result in rejection of the material back to the supplier. The supplier will be notified immediately by phone of the rejection, and follow-up with a Supplier Corrective Action Report (SCAR) form will occur within 12 hours. If the incident is a third nonconformance, the supplier will fall into rejection status and cannot be used for future purchases. Results that are out of specification for spoilage organisms will be accepted for use if they are to be used in a cook process provided the ingredient meets quality and regulatory (labeling, nutrient) requirements. Ingredients used in cold blend products will be rejected if out of specification. Notify the supplier immediately by phone and follow up with a SCAR form. If the incident is a third nonconformance, the supplier will be contacted by purchasing and QA to determine the source and correction of the problem.

   c. Require that R&D source ingredients only from approved suppliers, including ingredients for plant test runs.

4. Annually re-evaluate ingredient status to determine the need to change categories. Ingredients may change categories as history or use factors change.

Implementation of a thorough ingredient evaluation process can help prevent bringing contaminated materials into the plant and allow the plant to determine how they will be handled to prevent further contamination or growth of inherent organisms.

Sanitary Design and Materials

Sanitary design of the facility and equipment is covered in greater detail in Chapter 6. This is an important part of the microbiological control process in that it makes it easier to clean equipment, helps keep it clean during production, and prevents conditions that can contribute to product contamination.

In addition to equipment design to control microorganisms, there are materials available to industry that provide an added measure of microbiological control. These include equipment using antimicrobial stainless steel material that contains silver ions to both control surface bacteria and prevent the formation of biofilms and antimicrobial belting material. In addition, there are antimicrobial flooring materials. It should be understood that the use of these materials is part of a multiple-hurdle approach that includes sanitation as the primary means of control. These surfaces must still be cleaned for them to be effective. If the surface is not clean or if films form, bacteria to be controlled will not contact the antimicrobial surface. Another recent innovation is the use of the antimicrobial agent chlorine dioxide in packaging film to provide extended lethality. It may be used for meat and poultry, seafood, and fruits and vegetables [23].

Lubricants are used on food plant processing equipment to protect metal equipment from excessive wear. In the process of equipment movement, the lubricant may come in contact with the product, especially depending on the lubrication point's proximity to the product stream and the amount of lubricant applied. For this reason, the lubricants used must be food-grade. In the event that the lubricant does make its way into the product, it is limited by the FDA to 10 PPM [8]. If non-food-grade lubricants are used, the FDA has a zero tolerance for contamination. The former USDA, now National Sanitation Foundation International, designation for food-grade lubricant is H1. In addition to using food-grade lubricants as part of a multiple hurdle strategy for microorganism control, the use of lubricants with antimicrobial agents, such as sodium benzoate, is recommended. These agents either control the microorganisms through inhibition of further growth or with knockdown capability [9].

Sanitation

Food plant sanitation is covered in greater detail in Chapter 7; however, sanitation of the facility and equipment cannot be overemphasized. Starting the production day with clean equipment and maintaining sanitary conditions during operations will help protect product from disease-causing bacteria and spoilage organisms, harmful chemicals (i.e., allergens), and foreign material. Effective sanitation will also aid in the prevention of the formation of biofilms. Biofilm formation and control will be presented in Chapter 5.

Microbiological Testing and Validation

Microbiological testing of the food manufacturing environment can be a valuable tool to ensuring product safety. Environmental testing as verification of sanitation will be covered in greater detail in Chapter 8. It is important to emphasize that microbiological testing must be based on sound science and not random testing for the sake of testing. The plant must use the data generated to evaluate the effectiveness of the microbiological control strategies implemented.

Good Manufacturing Practices

Once the plant is clean it is very important that employees and visitors follow basic Good Manufacturing Practices (GMPs) to prevent contamination of clean surfaces. Though GMPs can prevent chemical and physical contamination of product, they can be most effective for microbiological control. The focus from a microbiological control standpoint will be on raw and cooked separation, hand washing, employee dress, and disease control. Basic GMPs are covered in greater detail in Chapter 9.

Pest Control

Pests, such as rodents, insects, and birds, can be a source of microbiological contamination in the food plant through their excreta and the parasites they carry. As part of an effective microbiological program, pests must be excluded and eliminated. Standards for an effective Pest Control Program are covered in greater detail in Chapter 10.

Employee Education

The importance of educating and training plant employees cannot be overestimated with regard to microbiological control. All plant personnel must understand their role in preventing microbiological contamination of products. As with many plant programs, this begins with the plant manager, who must be relied upon to support the microcontrol programs originating from corporate or plant QA. These include the purchasing group buying from approved suppliers and the engineering/maintenance department employing sanitary design features into the facility and the equipment. The sanitation department must understand its critical role in eliminating microorganisms through standard sanitation practices, and all employees must participate using GMPs to prevent bacterial contamination.

An understanding of microorganisms, their growth needs, and their control is an important first step in preventing their entry into food products. Prevention of contamination is the most important step in production of safe, wholesome food products and is much more effective than detection of microorganisms in finished products.

References

1. Ryser, Elliot T. and Elmer H. Marth, *Listeria, Listeriosis and Food Safety*, Marcel Dekker, New York, 1999, pp.
2. Frank, Hanns K., *Dictionary of Food Microbiology*, Technomic Publishing Company, Lancaster, 1992, pp.
3. Stauffer, John E., *Quality Assurance of Food, Ingredients, Processing and Distribution*, Food and Nutrition Press, Westport, 1994, pp.
4. Gould, Wilbur A., *CGMP's/Food Plant Sanitation*, CTI Publications, Baltimore, 1994, pp.
5. Christian, J. B. H., et al, *Micro-Organisms in Foods 2*, University of Toronto Press, Toronto, 1986, pp.
6. Troller, John A., *Sanitation for Food Processing*, Academic Press, 1983, chap. 6.
7. Zotolla, Dr. Edmund A., *Introduction to Meat Microbiology*, AMI Center for Continuing Education, 1972, pp. 8–9.
8. Hodson, Debbie, Food-Grade Lubricants Can Make Your Product Safer, *Food Quality*, 2004, pp. 90–92.
9. Anon, Improved Anti-microbial Agents in Food-Grade Lubricants, *Food Quality*, 2004, p. 96.
10. Imholte, Thomas J. and Tammy K. Imholte-Tauscher, *Engineering for Food Safety and Sanitation*, Technical Institute of Food Safety, Medfield, 1999, pp. 4–6.
11. Katsuyama, Allen M., *Principles of Food Processing Sanitation*, Food Processors Institute, 1993, p. 65, 77.
12. Tompkin, R. Bruce, Virginia N. Scott, Dane T. Bernard, William H. Sveum, and Kathy Sullivan Gombas, Guidelines to prevent post-processing contamination from Listeria monocytogenes, *Dairy Food and Environmental Sanitation*, 1999, pp. 551–562.
13. Henning, William R. and Catherine Cutter, *Controlling Listeria monocytogenes in Small and Very Small Meat and Poultry Plants*, Washington, D.C., pp.

14. Tompkin, R. B., Control of *Listeria monocytogenes* in the food processing environment, *Journal of Food Protection*, 2002, pp.
15. Ray, Bibek, *Fundamental Food Microbiology*, CRC Press, Boca Raton, 1996, pp. 15–17, 141–143.
16. Anon, *Bacteria, Yeast and Mold*, Extension Food Science Department, Cooperative Extension Services, University of Georgia.
17. Jay, James, *Modern Food Microbiology*, D. Van Nostrand Co., 1984, p. 295.
18. Herschdoerfer, S. M., *Quality Control in the Food Industry, Volume 1*, Academic Press, London, 1984, pp. 87, 89, 95, 98.
19. Dennis Buege and Steve Ingham, *Process Validation Workshop*, presented at AAMP Convention, Lancaster, PA, August 2000.
20. Mustapha, A. and M. B. Liewen, Destruction of *Listeria monocytogenes* by sodium hypochlorite and quaternary ammonium sanitizers, *Journal of Food Protection*, Vol. 52, No. 5, 1989, p. 306.
21. Anon, *Guidelines to Prevent Post-processing Contamination from* Listeria monocytogenes, National Food Processors Association, April 1999.
22. Anon, Guidelines for Developing Good Manufacturing Practices (GMPs), Standard Operating Procedures (SOPs) and Environmental Sampling/Testing Recommendations (ESTRs) Ready To Eat (RTE) Products, North American Meat Processors; Central States Meat Association; South Eastern Meat Association; Southwest Meat Association; Food Marketing Institute; National Meat Association; American Association of Meat Processors, 1999.
23. Bricher, Julie, Innovations in Microbial Interventions, *Food Safety Magazine*, The Target Group, Glendale, 2005, p. 33.
24. de Vries, John, *Food Safety and Toxicity*, CRC Press, Boca Raton, 1997, pp. 25–27.
25. Gombas, Dr. Dave, Environmental Monitoring for *Listeria*: Getting Started, *Food Safety Magazine*, April/May 2012, pp. 26–27.

Chapter 4

Control of *Listeria* in Food Manufacturing

Don't worry, *Listeria* floats in water so when we rinse off the equipment the *Listeria* will just float away and down the drain!

—Anonymous food plant employee

Listeria

Listeria is a ubiquitous organism that can be found almost everywhere in nature. There are eight identified species of *Listeria* (*dentrificans, grayi, innocua, ivanovii, murrayi, seeligeri, welshimeri,* and *monocytogenes*); however, only one species, *Listeria monocytogenes*, has been determined to be a human pathogen. The severity of *L. monocytogenes* in foods and listeriosis, the disease it causes, has been well documented. The first report of foodborne illness from *L. monocytogenes* was in Canada in 1981. About 41 people became ill and 18 died after consuming contaminated coleslaw [20]. One of the earliest cases of listeriosis investigated in the United States involved contaminated milk in Massachusetts, infecting 49 people and resulting in the death of 14 [20]. A second *L. monocytogenes*–related foodborne illness in the United Stages involved Mexican-style soft cheese in Southern California in 1985. The presence of *L. monocytogenes* in the cheese resulted in the death of 39 people [1]. Subsequent to these cases, there have been well publicized cases of listeriosis and product recall in various Ready To Eat (RTE) meat and poultry products. Some of these cases have led to severe illness and death, whereas others led to very expensive product recall. It was the leading cause of food recalls in 1999–2000 and, as a result, there is a zero tolerance for *L. monocytogenes* in RTE foods, and its presence in RTE foods is considered adulteration, and product in distribution is subject to recall or seizure.

Fortunately, the incidence of listeriosis has declined as a result of the food industry taking significant action to reduce the entry of the organism into manufacturing plants, improvement in sanitation practices, and prevention of post-lethality product contamination. Much of this is due to the large numbers of recalls in the late 1980s through the 1990s and the attention that the organism and the disease it causes received by the industry and regulatory agencies. As a result of the actions taken, the rate of infection decline is approaching the goal of 50% reduction to less than 2.5 million cases in 2005 [24]. Because *Listeria* is a ubiquitous organism, meaning that it is found in many areas of the environment, it is unlikely that manufacturers will be able to completely eliminate it from the manufacturing environment; however, through specific steps they can manage and control *L. monocytogenes*, thus preventing its entry into their food products.

Requirements

The requirements of the organism are identified in greater detail in Chapter 3. *Listeria* is a Gram-positive rod-shaped bacillus, non-spore forming and motile. It is facultative, meaning that it can live in an aerobic (with oxygen) environment or adapt to an anaerobic (without oxygen) environment. It is often referred to as a ubiquitous organism, meaning it can be found frequently throughout the environment. *Listeria* is commonly found in humans, soil, animal feces, raw vegetables, and cheeses made from raw milk. There is a high incidence rate in raw meat and poultry. The pathogenic form, *L. monocytogenes*, is readily found in raw meats, anywhere from 16% to 92% in various studies. It is more heat resistant than other pathogens but is destroyed by cooking. However, it is particularly challenging to control in food production because it can grow in a wide range of conditions. The organism grows in the temperature range 32°F–113°F and survives freezing, although it stops growing below 31°F. *Listeria* can grow in the pH range 5.2–9.6, is salt tolerant between 5% and 10%, and has a water activity (Aw) greater than 0.93.

Transmission of *Listeria* (*Listeria* implies *Listeria monocytogenes* for the rest of this chapter) includes raw material, biofilms on equipment, aerosols, the environment, and people (7%–25%). Kill steps for *Listeria* are sufficient thermal application for lethality as indicated in USDA Appendix A, and the application of temperature over time needs to be sufficient for a 5–7 log reduction to provide a good safety margin. *Listeria* is less heat resistant than *Salmonella*; however, both have increased tolerances in a high-fat system. It is post-lethality contamination that has posed a problem for the food manufacturing industry, particularly in RTE meat and poultry products. In these products, *Listeria* is considered to be an adulterant, meaning that the product is subject to disposal by the producer or recall if the product is in distribution.

Listeriosis

The disease in humans that is caused by *L. monocytogenes* is called listeriosis. As indicated before, the incidence of listeriosis has declined in the last several years through the actions of the food industry. The decline is also due in part to the actions of the federal government, from regulations and requirements by the USDA to disease monitoring by the Centers for Disease Control and Prevention (CDC). Surveillance of listeriosis by CDC occurs through PulseNet and Food Net. These two systems facilitated information sharing between public health laboratories on the genetic fingerprints of strains isolated from all over the country [26]. Through this information sharing, the labs were able to differentiate between isolated incidents of listeriosis and possible outbreaks with a common source. This then provided a scientific tool to identify situations where adulterated product should be removed from the consumer market through recall. It also provides the ability to connect product from an outbreak to the specific source by matching the genetic identity. This means that the manufacturing plant can swab the facility and fingerprint any swab samples that test positive for *L. monocytogenes*. If any of the fingerprints match that of the genetic material from the adulterated product, the plant has a better idea of the source and can focus their attention on elimination or control of the source.

As indicated in Chapter 3, foodborne pathogenic organisms cause illnesses in two different ways. Listeriosis occurs through bacterial infection when there is ingestion of food adulterated with live bacteria. *Listeriosis monocytogenes* has a relatively high infectious dose in healthy individuals, but it has a relatively low infectious dose, possibly less than 1000 cfu/g [27], in "at-risk" or susceptible individuals [19]. Susceptible individuals are the very young and the very old, people with compromised immune systems (i.e., those with AIDS or cancer patients undergoing chemotherapy), and pregnant women. Symptoms of the disease include fever, chills, abdominal pain and nausea, diarrhea, and headache. As the disease progresses, it may mimic meningitis and can result in a toxic blood condition called septicemia. What is most hideous about the disease is the impact it has on pregnant women because it can cross the placental barrier, putting the fetus at risk, and the infection can result in spontaneous abortion or stillbirth. Onset of the disease can be between 7 and 30 days from initial exposure, which may make it difficult to identify the specific source of infection.

Listeria Harborage and Contamination

Because of the ubiquitous nature of *Listeria,* it is continually introduced into food manufacturing environments [14]. For this reason, *Listeria* management, especially in RTE food operations, involves a recognition of areas where *Listeria* can harbor and grow.

Contamination occurs when fully cooked product is exposed to a contaminated surface. The surface may become contaminated from a bacterial growth niche. Growth niches are locations where the organism grows and multiplies [14]. A niche is also where the organism is found after flood sanitizing. These niches may not be the transfer point of contamination but may contaminate the transfer point; thus, they must be designed out of the process [24]. This may be accomplished through sanitary design and must be managed through additional management strategies. *Listeria* is most often found in the following locations in an RTE food manufacturing plant in order of their frequency of occurrence: floors, drains, cleaning aids, wash areas, sausage peelers, food contact surfaces, condensate, walls and ceilings, and compressed air [23]. Post-lethality contamination comes from these areas, and they require implementation of control measures to eliminate the organism and prevent reestablishment in the niche area.

Listeria Management and Control Methods

One of the primary factors contributing to the adulteration or product in an RTE environment is the formation of a growth niche. Contributing factors to growth niches are equipment design problems, product debris working into unclean locations, mid-shift cleaning, and the use of high pressure during cleaning and procedures requiring excess moisture [24]. Control of microorganisms in a food manufacturing operation requires many steps, which is often referred to as a "multiple hurdle" approach. This means that many strategies will be employed primarily to prevent microorganisms from having an opportunity to come into the facility, establish harborage in the facility, or grow if they become established. It also involves elimination of the organism from ingredients and prevention of post-lethality process recontamination. There are formulating measures that can be followed to affect the pH; moisture level, especially water activity; salt level; and inhibitory ingredients such as sodium nitrite, potassium lactate, and sodium diacetate can also be used. In addition, there are packaging strategies such as vacuum packaging, gas back flush, and the use of oxygen scavengers to prevent growth of bacteria that may be introduced post lethality. In addition, there are post-packaging treatments such as hot water, steam, or pressure post pasteurization that are effective in destroying bacteria present in the food package. The use of gamma irradiation is the most effective post-pasteurization process along with x-ray. Both are extremely expensive, impractical for in-plant application, and not approved for compound meat and poultry products at this time. E-beam is not as effective as gamma or x-ray, but may be more practical for in-plant application. These measures are all effective and part of a comprehensive food safety system. To some extent, these measures assume that bacteria are present in the post-lethality finished product. A more practical and cost-effective food plant strategy is to prevent post-lethality product contamination so that post-lethality treatments are not required.

Control of *Listeria* requires food manufacturers to make a concerted effort to eliminate niches through improved equipment and process design, validation of processes, and management [24]. This section will focus on some of the measures to be employed by food facilities to enhance sanitation and prevent contamination of food products post-lethality and into packaging.

One of the most effective management strategies to prevent post-lethality recontamination is a "clean room" process for exposed RTE product. Clean rooms must be a controlled environment (traffic, air HEPA filtered, etc.), validated microbiologically, approach pharmaceutical standards of cleanliness, and disinfected. Disinfection involves application of sanitizer after cleaning at levels of 1000–1200 ppm for the complete removal of pathogens. Another approach is to create mini-environments: critical areas in the process where the product is contained and post-lethality process contamination is prevented. In all cases, an aggressive control strategy is required. The following strategies may be applied to those areas previously identified as having the highest frequency of findings in the RTE plant environment.

Floors and Drains: A dry environment is preferred, especially for floors and drains, as *Listeria* requires water for growth and survival and it is often transported by droplets of water aerosolized from high-pressure spraying on floors or in drains. Specific cleaning procedures for floors and drains, identified in Chapter 7, will be applied to eliminate these areas as growth niche sources. People cleaning floors and drains should not clean RTE equipment or should clean the floors and drains after cleaning production equipment. Floors that are wet due to normal plant operating conditions will be kept sanitized (Figure 4.1).

Floors must be in good condition, with no cracks and no low spots to allow pooling water. Sanitize the entire floor; don't just rely on run-off from equipment. Eliminate open trench drains, and place quat blocks in drains to control *Listeria* growth after cleaning and sanitizing. Have a procedure in place for controlling and cleaning floors and drains after a drain backup as presented in Table 4.1.

Plant layout and traffic control are also critical to the prevention of transfer of *Listeria* on the floors. Control movement of people, equipment, and forklifts and pallet jacks to prevent them from traveling from raw to RTE product areas. Where possible, separate wheeled vehicles between raw and RTE areas (Figure 4.2).

When wheeled trash or inedible material containers are used, they must be cleaned frequently. Where there are common areas for foot or wheeled traffic, use floor mats or sanitizer spray foamers to sanitize shoes or wheels passing through these areas.

Figure 4.1 Floor foamers are valuable in areas where the floor is usually wet. Foot and wheeled traffic pass through the sanitizer foam to prevent spread of contaminants to RTE areas, especially where there are common traffic areas.

Cleaning aids: The sanitation staff must have clean and sanitized gear (i.e., boots, raincoats, and aprons) that can be stored in a location that will allow drying and prevent contamination. Clean and sanitize Personal Protective Equipment and cleaning tools such as brooms, scrubbing brushes, squeegees, and floor scrubbers at a level of 600–800 ppm quat. Identify cleaning tools by color and specific use. For example, cleaning tools used in raw areas may be identified by the color red, whereas tools used for RTE cleaning may be identified by the color blue. These tools will be stored apart to prevent cross-contamination. In addition, tools used to clean floors and drains will be stored apart from tools used for cleaning of equipment surfaces. There should not be any wooden-handled cleaning tools, such as brooms or shovels. Eliminate wood from as many locations as possible in the food plant, especially in the RTE product areas. Do not use cloth mops for sanitation purposes.

Wash areas: The plant should have separate wash areas for raw and RTE equipment. Using one area for all equipment may result in cross-contamination of RTE equipment by the raw equipment or conditions in shared wash rooms. Washrooms must have good ventilation to minimize for and condensate, but should have a negative air pressure. It is a good idea to swab washrooms periodically to ensure that they are clean and do not contribute to equipment contamination.

Sausage peelers: Hotdog and sausage manufacturers have a challenge with peelers as there is constant re-exposure to food, and they are hard to clean due to

Table 4.1   Drain Backup Procedure

Cleo's Foods Sanitation Performance Standard Operating Procedure		
Procedure Number: 2	Version: 2 Dated: 05-13-05	Replaces Version: 1 Dated: 06-26-04
Procedure: Drain Backup Corrective Action		Pages: 2

A. OBJECTIVE: To prevent bacteria from spreading through the plant due to drain backup onto floors. This procedure will pertain to all frozen, ready-to-eat, and raw areas.

B. RESPONSIBILITY: Plant Employees

C. EQUIPMENT: Quaternary ammonia (800–1000 ppm) solution, squeegee, brushes or brooms, drain equipment.

D. FREQUENCY: As needed.

E. PROCEDURE: All personnel, equipment, ingredients, post-lethality products, etc., are to be moved away from the backed-up drain area prior to cleanup. No one is allowed to walk through this area except maintenance and/ or sanitation associates correcting the problem.

Footbaths will be stationed at all places of ingress and egress of the unsanitary area.

Note: Everyone is required to use the footbaths and change smocks upon entering and leaving the unsanitary area.

The area will be sectioned off and QA Hold tags positioned to notify employees not to enter.

All raw and ready-to-eat equipment and materials need to be moved or covered prior to unplugging the drain.

Maintenance is to be notified to determine the severity of the drain backup, which may require mechanical equipment or calling an outside service specialist.

Again, *no* personnel, except sanitation, maintenance, or outside service personnel, are to enter the area being serviced.

Once the drain is unplugged, Sanitation will commence cleaning and sanitizing the areas.

Floors and the drain need to be scrubbed using soap prior to the application of quat solution.

Squeegee is to be used to remove excess water from the floor.

Quat or a similar approved sanitizer is to be used to sanitize the floor.

Continued

Table 4.1 (*Continued*)   Drain Backup Procedure

Cleo's Foods *Sanitation Performance Standard Operating Procedure*
Once all work and/or cleanup of the drain backup has occurred, Sanitation and/or Maintenance will change smocks before entering the unaffected areas. Personnel and materials will re-enter the room once QA releases it. Confidential Commercial Information

Figure 4.2   Separation of equipment would preclude situations where jacks designated for cooked areas only are used to handle raw meat materials.

many small and moving parts, especially older models [24]. For thorough cleaning purposes, they must be broken down on a nightly basis. They may require regular baking in a smokehouse or application of steam for deep cleaning. If either method is used, the surface temperature must reach 160°F for 20 to 30 min. Application of quat at levels of 200 ppm periodically during production to the peeler and pans may be necessary to maintain sanitary conditions.

Food contact surfaces: Do not use the same equipment for raw as for cooked, even with cleaning. Seals and gaskets from RTE product equipment must be removed frequently for cleaning and sanitizing as they can be harborage areas for bacteria. They should be evaluated for damage and placed on a preventive maintenance schedule for replacement before they create a microbiological or physical

Figure 4.3   Placement of equipment parts on the floor, like this canister lid, can contribute to cross-contamination of RTE product or ingredients.

hazard. Avoid catwalks or stairs over food lines; however, if they are needed, they must have sealed legs and rails, no hollow areas and kick plates. Never place dismantled food equipment parts on the floor, especially if they are used for RTE product or ingredients (Figure 4.3).

Condensate: This can be a source of contamination if it comes off a contaminated source and falls onto product or product contact surfaces (Figure 4.4).

Walls and ceilings, refrigeration units: These units require cleaning of fins, coils, and pans on a weekly basis. The use of quat blocks in the drip pans will reduce the chances that moisture from the pan will become contaminated and contaminate product if it drips onto product or contact surfaces. Drains from the drip pans should run directly to floor drains and not onto floors.

Compressed Air: Compressed air will be filtered for 99.99% efficiency at 0.2 microns and dehumidified to moisture in the lines.

In addition to the procedures listed above, extra precautions are needed to prevent contamination of RTE products. Listed below are additional control measures that can be applied to control *Listeria* in the plant to prevent product contamination.

Ingredients

Ingredients used in food manufacturing can be a source of microbiological contamination and the hazards vary by the ingredient source. *Listeria* can often be

Figure 4.4 Where condensate cannot be completely controlled, such as on the outside of freezers, shields can be installed to channel moisture away from product or product surfaces.

found in raw agricultural products and has a very high association with meat and poultry raw materials, or raw materials from other animal sources [22]. Testing of finished product for spoilage organisms or pathogens is only one tool in verifying quality or safety; however, it is not an effective means of ensuring quality and safety as the sample size required for statistical significance is usually quite high. The focus of microbiological analysis should be on ingredient components, rather than on finished product, to ensure that ingredient supplier processes are in control and that product contamination from ingredients is prevented. Using ingredient hazard analysis, a company can set up microbiological risk categories for each of the ingredients that they use in their operation. All ingredients will be identified for levels of microbiological sensitivity based on potential hazard, history, composition, ingredient source, and the product/process into which the ingredient is used. Once this is done, they can establish the requirements for supplier testing and Certificate of Analysis (COA) information as well as internal needs to evaluate incoming raw materials. This will include the frequency with which the ingredients are analyzed, the analysis required, and how the results will be used. Ingredients will be tested to verify that they will not contribute to product contamination and to verify that supplier processes provide conforming ingredients. Disposition of nonconforming ingredients and identification of nonconforming suppliers will be dealt with in the procedure. This procedure was presented in greater detail in Chapter 3; however, it has important applications in the prevention of contamination by *L. monocytogenes*.

Plants bringing in fully cooked meat and poultry chubs that will go into RTE products that receive no further lethality should sanitize the outside of the chub with 200 ppm quat before opening as a means of precautions [23].

Sanitary Design and Materials

Unfortunately, many plants currently manufacturing RTE meat and poultry product or other potentially sensitive products were not designed or built with control of *L. monocytogenes* in mind. The same is true of equipment used for these types of products. Consequently, *Listeria* is able to enter the facility through raw materials, people, or equipment, establish itself in the environment and grow, increasing the potential of product contamination. A very important part of management and control is sanitary design of the facility and equipment. This is covered in greater detail in Chapter 6; however, it must be understood that this is a key component to reducing the exposure and incidence of contamination. The primary measures of control through sanitary design are

- Creation of sanitary zones within the plant and controlling traffic between the zones to prevent exposure in RTE areas. Potential crossover between raw and cooked in oven areas if the ovens are not flow through. Separate entrance and lunch areas for raw and RTE employees.
- Use of materials for the plant and the equipment that are sturdy under normal operating conditions, smooth and nonporous, and easily cleanable. Structural materials should not contribute to product or product surface contamination (Figure 4.5).
- Creation of effective airflow, high pressure in cooked product rooms to low pressure in raw auxiliary production areas. Ventilation to prevent condensation and filtration of incoming and compressed air so that these do not contribute to product contamination. Refrigeration of plant air is beneficial to reducing microbiological growth as Listeria grows twice as fast at 50°F as it does at 40°F [23].
- Sanitation. This is one of the most important parts of the microbiological control processes for *Listeria* in that sanitary design makes cleaning of facility and equipment easier, helps keep it clean during production, and prevents conditions that can contribute to product contamination.

In addition to equipment design to control microorganisms, there are materials available to industry that provide an added measure of microbiological control. These include equipment using antimicrobial stainless steel material and antimicrobial belting material. In addition, there are antimicrobial flooring materials and antimicrobial lubricants. It should be understood that use of these materials is part of a multiple-hurdle approach that includes sanitation as the primary means of

Figure 4.5 Ceilings and overhead structures may contribute to contamination of product or production surfaces if they are not smooth, nonporous, and free of dust and soil-collecting pipes or conduit.

control. These surfaces must still be cleaned for their antimicrobial properties to be effective.

Lubricants are used on food plant processing equipment to protect metal equipment from excessive wear. In the process of equipment or equipment part movement, the lubricant may come in contact with the product, especially depending on the lubrication point proximity to the product stream and the amount of lubricant applied. For this reason, the lubricants used must be food-grade. In the event that the lubricant does make its way into the product, there is an FDA limit of 10 ppm [8]. If non-food-grade lubricants are used, the FDA has zero tolerance for contamination. Previously, USDA approved and designated chemicals as food-grade; now, however, the National Sanitation Foundation International provides the designation for incidental contact food-grade lubricant as H1. These lubricants are formulated with ingredients listed in 21CFR178.3570 [25]. As part of a multiple hurdle strategy for microorganism control, the use of lubricants with antimicrobial agents, such as sodium benzoate, is recommended. These agents either control the microorganisms through inhibition of further growth or with knockdown capability [9]. Lubricants with antimicrobial ingredients must be registered with the US EPA, comply with FDA regulations, and have Generally Regarded As Safe (GRAS) status [25].

Sanitation

Food plant sanitation is covered in greater detail in Chapter 7; however, sanitation of the facility and equipment cannot be overemphasized. Starting the production day with clean equipment and maintaining sanitary conditions during operations will help protect product from disease-causing and spoilage organisms, harmful chemicals (i.e., allergens), and foreign material. Effective sanitation will also aid in the prevention of the formation of biofilms. Biofilm formation and control will be presented in Chapter 5. Cleaning and sanitizing procedures should be focused on the control of *Listeria* and may include specific processes as follows [14,21]:

- Special training will be required for sanitarians in control methods for *Listeria*. They need to know the extent to which equipment needs to be broken down for cleaning [24]. Hoses must not draped across equipment, and the spray heads must be kept off the floor. Remove hoses from production areas after sanitation is completed or cover them with plastic. Clean and swab hoses regularly to verify that they are clean.
- Outer protective gear for the cleaning crew will be washed, sanitized, and dried after use. It will be stored in a manner that will maintain sanitary conditions. Even so, sanitarians shall not lean on clean equipment with their protective gear.
- Coolers should be cleaned on a regular basis, especially if they are used for storage of hotdogs, sausages, or luncheon meat logs for slicing. Do not clean coolers when product is present, even if it is shrouded in plastic, and do not spray floors with water as this may create aerosols that can contaminate product. Large exposed product freezers, such as spiral freezers, require thawing before cleaning. Repeated thawing, cleaning, and refreezing may result in structural damage and creation of bacterial harborage niches. It is better to clean freezers less frequently; however, this decision must be supported by documentation.
- Items that are often handled by plant personnel or equipment operators must be included on a daily cleaning schedule. This includes pull cords or push buttons used to open rollup or sliding doors as these are handled by many people from several areas of the plant. Push-in control buttons present a special potential for microbiological harborage and must be removed periodically for cleaning or replaced with the "mushroom"-style button. Clean equipment touch screens by wiping them with a mild alcohol solution.
- Special inspection and cleaning may be required for equipment brought in from outside the plant or from an outside storage area. Equipment that has been stored outside must be thoroughly inspected for damage or insect infestation. It should be broken down for cleaning and sanitizing prior to use.

Product

Specific procedures may be required for production to ensure the safety of products being made, especially if these are fully cooked RTE meat and poultry products. One of the most important procedures is verification that the product has reached lethality as specified in FSIS Appendix A, which stipulates the times at temperatures needed to achieve sufficient log reduction of microorganisms. Verification that product has reached the temperature specified at the Critical Control Point in the plant HACCP plan is important as the plant does not want to package and ship undercooked product. Other processes that may be required are

- Validated cooking procedures. Scientific validation of the plant cooking process is an important means of ensuring that the process will achieve a 6- to 7-log reduction of pathogens. Validation is best conducted at a third-party lab with expertise in this area. The process involves inoculating raw ingredients at high levels of pathogens, usually *L. monocytogenes*, *Salmonella*, and in ground beef products, *E.coli* O157:H7. For this reason, validation studies should never be conducted in the manufacturing plant. The inoculated ingredient is heated following the plant cooking process, and the cooked product is tested for the specific pathogens to verify the log reduction.
- Stipulation to all employees that any exposed RTE product touching the floor will be discarded with no exceptions.
- Specific procedures for handling rework. Product should be taken back through a lethality process unless it only requires repackaging (i.e., for leaker hotdog packages). Product should only be repacked under controlled conditions where there is assurance that the product has not been contaminated. There should be a procedure for sanitizing outside of packages of product touching the floor if they are not damaged or subject to rework or repackaging. Product that has been returned from a customer must be thoroughly evaluated to assess their condition before being reworked or reprocessed.

Microbiological Testing and Validation

Microbiological testing of the food manufacturing environment can be a valuable tool to ensuring product safety. Environmental testing will be covered in greater detail in Chapter 8. It is important to emphasize that microbiological testing must be based on sound science and not random testing for the sake of testing. The plant must use the data generated to evaluate the effectiveness of the microbiological control strategies implemented. Map the environmental swab results to show high or hard-to-control areas in the plant to develop control strategies [14]. Studies have revealed seasonality with higher prevalence in the summer as well as in different climates [14].

Personnel

New food plant employees may not be familiar with the needs for microorganism control in the food manufacturing environment, and therefore training is critical [23]. Though GMPs can prevent chemical and physical contamination of product, they can be most effective for microbiological control. The focus from a microbiological control standpoint will be on management commitment, raw and cooked separation, product handling, and employee dress. Basic GMPs are covered in greater detail in Chapter 9, and these procedures will help prevent product contamination from employees in a post-sanitation and post-lethality environment. However, the following are additional considerations for control of *Listeria* in the environment and prevention of cross-contamination:

- Management must set the example for all plant employees to follow GMPs and all food safety processes related to prevention of microbiological contamination. By establishing food safety as a priority, management can demonstrate personal responsibility to all employees and visitors. This includes following requirements for wearing the correct color smock, washing hands, and following plant flow requirements from cooked to raw when taking visitors through the facility.
- Mechanical personnel have a significant impact on food safety as they are often required to conduct setup and repair on equipment, specifically in RTE areas. In order to prevent contamination from equipment setup and repairs, mechanics must be expected to follow all GMPs. When possible, they should have separate tools for use in raw and RTE areas. When this is not possible, the tools will be sanitized with an alcohol solution to prevent rusting. Sanitary design for tools will eliminate potential niches (Figures 4.6 and 4.7).
- Personnel responsible for spill and trash control or control of condensate must be trained not to handle ingredients, product, or packaging. They must not be allowed to work on production lines unless they have gone through a smock change and thorough hand washing and sanitizing.

Pest Control

Pests, such as rodents, insects, and birds, can be a source of microbiological contamination in the food plant through their excreta and the parasites they carry. As part of an effective microbiological program, pests must be excluded and eliminated. Standards for an effective pest control program are covered in greater detail in Chapter 10.

Construction

Construction provides opportunities for microorganisms, specifically *L. monocytogenes* that have been harbored in damp areas inside walls or under floor coatings, to

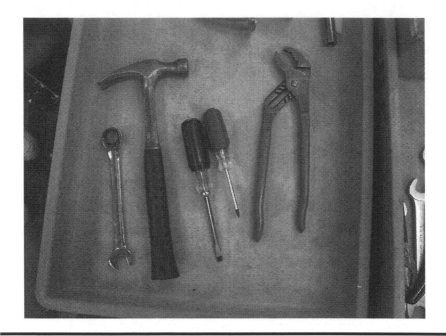

Figure 4.6 Sanitary design of mechanical tools will eliminate plastic coatings that can create bacterial harborage. Tools will be maintained in a clean condition and will be in cleanable tool boxes or on cleanable carts.

be redistributed in the facility. This is particularly true when construction involves demolition of walls, digging floors to replace drains, or removal of damaged floor coating. For this reason, additional procedures must be taken when demolition and construction are going on in the plant. While organisms like *Listeria* are not typically airborne, they can be carried on particles of dust raised during construction, they can be moved by the contractor and maintenance traffic patterns, and they can be translocated by removal of demolition materials. Procedures that a plant can employ involve suppression of dust, prevention of cross-contamination, and verification of control. Suppression of dust can be accomplished by creating negative pressure in the room where construction in being conducted. However, the best method for preventing demolition/construction dust from spreading potential contamination through the plant is by the construction of temporary walls to separate work areas from production areas (Figure 4.8).

There are several steps involved in the creation of construction barriers, or containment, starting with decisions about the need for containment and the type of containment that can be created. First, the plant management team meets to review the construction/demolition project so that all factors can be weighed. The team will consist of the Plant Manager, QA Manager, Maintenance Manager, and Sanitation Manager at a minimum. The Maintenance Manager will describe the scope of the project, what construction will be involved, whether contractors will be used, and

Figure 4.7   Leather tool pouches are absorbent and not able to be cleaned and sanitized. They can be a source of contamination and should be eliminated in favor of cleanable tool carriers.

the duration of the project. Second, the team will determine whether production must be stopped or if it can continue with the construction area securely separated from ingredients, production equipment, or product. Third, the team will then determine the type of containment that must be created. Containment can be as simple as heavy gauge plastic sheeting taped to walls or as durable as a wooden frame covered with heavy gauge plastic. If a simple containment can be created, then it must be secured to walls, the ceiling, and the floor so that no dust or contaminated air can escape from the construction area. When a wooden frame is constructed, it must also create a complete seal against walls and floors as well as overheads.

Fourth, the team must determine the traffic pattern for maintenance and contractor personnel who will be working in or exiting from the containment area. The traffic pattern that will have the minimum amount of impact on the remaining environment will be established and will include the path by which demolition material is removed from the containment. If contractors are used, they must be trained to know and follow the GMPs that are required by the plant. The training must be documented, and they must be instructed not to enter areas of the plant that are involved in production.

Fifth, when the containment is built, it should be approved by the QA Manager. The approval will include an evaluation of the structure to ensure that it completely contains the area of demolition and construction, that it is secure to prevent airborne

Figure 4.8 Construction of barrier walls separating construction areas from production areas will help prevent the spread of microorganisms on construction dust.

contaminant from escaping above, below, or around the structure, and that objects inside the containment that cannot be washed (electrical panels, junction boxes) are covered so that they don't become contaminant harborage areas. In addition, the QA Manager will ensure that there are quat crystals or powdered floor sanitizer around the outside of the containment as a supplemental barrier in the event that there is a breach at the base of the containment. Once the containment is approved, and the approval should be documented, the demolition and construction process can begin.

Because of the presence of microorganisms in the containment and the potential that they can be spread through dust particulate, it is highly recommended that the plant increase the numbers of environmental swabs. The swabs will be collected daily while construction is going on and will be collected around the base and the edges of containment. They will be analyzed for *Listeria* species. The analytical data from the swabs will aid the plant in verifying the effectiveness of the containment. Debris that is created from the demolition process will be collected in a specifically identified container that will be used to remove it from the containment. Before removal, the debris will be covered with a sheet of heavy gauge plastic that will be sealed around the top of the container. The plastic will be sprayed with 800–1000 ppm quat before removal from the containment and transport out of the plant.

Once the construction process has been completed and the containment is made ready for removal, it will be sprayed on the outside with 800–1000 ppm quat. It will then be folded in on itself and placed in the container used to remove demolition debris. The container will be covered with plastic, the plastic sealed, and then sprayed with quat at the 800–1000 ppm before removal. The construction area will then be cleaned and sanitized at least twice to ensure that no contaminant remains. Finally, environmental Listeria swabs will be colleted in the area, including from floors, walls, and overhead structures to verify that the cleaning and sanitizing process were effective.

Employee Education

The importance of educating and training plant employees cannot be overlooked with regard to microbiological control. All plant personnel must understand their role in preventing microbiological contamination of products. As with many plant programs, this begins with the plant manager, who must be relied upon to support the microcontrol programs originating from corporate or plant QA. There must be management commitment passed along to the plant employees [23]. It includes the purchasing group buying from approved suppliers and the engineering/maintenance department employing sanitary design features in the facility and the equipment. The sanitation department must understand their critical role in eliminating microorganisms through standard sanitation practices, and all employees must participate using GMPs to prevent bacterial contamination.

Control of *Listeria* involves several processes that must be employed by food manufacturing plants. These processes, when employed, will provide for the safety of the product and provide the company with security in knowing that they have employed all reasonable means of preventing product adulteration and consumer illness. An aggressive *Listeria* control program may also yield additional benefits from control of spoilage microorganisms, and improvement in product quality and shelf stability.

References

1. Ryser, Elliot T and Elmer H. Marth, *Listeria, Listeriosis and Food Safety*, Marcel Dekker, New York, 1999, pp.
2. Frank, Hanns K., *Dictionary of Food Microbiology*, Technomic Publishing Company, Inc., Lancaster, 1992, pp.
3. Stauffer, John E., *Quality Assurance of Food, Ingredients, Processing and Distribution*, Food and Nutrition Press, Inc., Westport, 1994, pp.
4. Gould, Wilbur A., *CGMP's/Food Plant Sanitation*, CTI Publications, Baltimore, 1994, pp.
5. Christian, J. B. H., et al., *Micro-Organisms in Foods 2*, University of Toronto Press, Toronto 1986, pp.
6. Troller, John A., *Sanitation for Food Processing*, Academic Press, 1983, chap. 6.
7. Zotolla, Dr. Edmund A. *Introduction to Meat Microbiology*, AMI Center for Continuing Education, 1972, pp. 8–9.

8. Hodson, Debbie, Food-Grade Lubricants Can Make Your Product Safer, *Food Quality*, 2004, pp. 90–92.

9. Anon, Improved Anti-microbial Agents in Food-Grade Lubricants, *Food Quality*, 2004, p. 96.

10. Imholte, Thomas J. and Tammy K. Imholte-Tauscher, *Engineering for Food Safety and Sanitation*, Technical Institute of Food Safety, Medfield, 1999, pp. 4–6.

11. Katsuyama, Allen M., *Principles of Food Processing Sanitation*, Food Processors Institute, 1993, p. 65, 77.

12. Tompkin, R. Bruce, Virginia N. Scott, Dane T. Bernard, William H. Sveum, and Kathy Sullivan Gombas, Guidelines to prevent post-processing contamination from *Listeria monocytogenes*, *Dairy Food and Environmental Sanitation*, 1999, pp. 551–562.

13. Henning, William R. and Catherine Cutter, *Controlling* Listeria monocytogenes *in Small and Very Small Meat and Poultry Plants*, Washington, D.C., pp.

14. Tompkin, R. B., Control of *Listeria monocytogenes* in the Food Processing Environment, *Journal of Food Protection*, 2002, pp.

15. Ray, Bibek, *Fundamental Food Microbiology*, CRC Press, Boca Raton, 1996, pp. 15–17, 141–143.

16. Anon, *Bacteria, Yeast and Mold*, Extension Food Science Department, Cooperative Extension Services, University of Georgia.

17. Jay, James, *Modern Food Microbiology*, D. Van Nostrand Co., 1984, p. 295.

18. Herschdoerfer, S. M., *Quality Control in the Food Industry, Volume 1*, Academic Press, London, 1984, p. 87, 89, 95, 98.

19. Dennis Buege and Steve Ingham, *Process Validation Workshop*, presented at AAMP Convention, Lancaster, PA, August 2000.

20. Mustapha, A. and M. B. Liewen, Destruction of *Listeria monocytogenes* by Sodium Hypochlorite and Quaternary Ammonium Sanitizers, *Journal of Food Protection*, Vol. 52, No. 5, 1989, p. 306.

21. Anon, *Guidelines to Prevent Post-Processing Contamination from* Listeria monocytogenes, National Food Processors Association, April 1999,

22. Anon, Guidelines for Developing Good Manufacturing Practices (GMP's), Standard Operating Procedures (SOP's) and Environmental Sampling/Testing Recommendations (ESTR's) Ready To Eat (RTE) Products, North American Meat Processors; Central States Meat Association; South Eastern Meat Association; Southwest Meat Association; Food Marketing Institute; National Meat Association; American Association of Meat Processors, 1999,

23. Anon, Interim Guidelines, Microbiological Control During the Production of Ready To Eat Meat and Poultry Products, Controlling the Incidence of Microbial Pathogens, Joint Task Force on Control of Microbial Pathogens in Ready To Eat Meat and Poultry Products, Washington, D.C., 1999,

24. Butts, John, *Seek & Destroy: Identifying and Controlling* Listeria monocytogenes *Growth Niches*, Food Safety Magazine, The Target Group, Glendale, 2003, pp. 24–29, 58.

25. Yano, Kenji, Advances in the Application of Food Grade Lubricants, *Food Safety Magazine*, The Target Group, Glendale, 2004, pp. 34–37, 65.

26. Donnelly, Catherine, Getting a Handle on *Listeria*, *Food Safety Magazine*, The Target Group, Glendale, 2003, pp. 18 – 27.

27. Anon, *Listeria Briefing: Requirements of FSIS Directive 10,240.3*, presented by the National Meat Association and The HACCP Consulting Group, Los Angeles, 2003.

Chapter 5

Biofilms

Formation

One of the challenges that face food processors and the sanitation team is the formation of biofilms on food equipment surfaces. Food manufacturing plants have recognized this potential hazard related to sanitation and that biofilms can have a profound impact on the safety and quality of their products. Their formation has the potential to contaminate product through the introduction of pathogenic microorganisms or spoilage bacteria. Biofilms have been described as "metabolically active matrix of cells and extracellular compounds" [6] or as "matrix enclosed bacterial populations adherent to each other and/or to surfaces or interfaces" [5]. They may contain spoilage bacteria such as *Pseudomonas fragi*, *Enterococcus* spp., and *Pseudomonas fluorescens* as well as pathogens such as *Listeria monocytogenes*, *Staphaylococcus aureus*, *Escherichia coli* O157:H7, or *Salmonella* [3]. They are difficult to remove as they are resistant to normal sanitation procedures, and they can result in other detrimental process effects. Even when a food surface appears to be clean, the presence of biofilms is a potential hazard that must be eliminated and prevented from reoccurring. Before this can be done, it is important to understand what a biofilm is and how it is formed.

Biofilms begin with a conditioning layer of organic (protein) or inorganic matter forming on an otherwise visually clean food contact surface. The accumulation of organic and inorganic material on processing surfaces creates an environment where bacteria can adhere. Bacterial adhesion is referred to as the conditioning layer that occurs when cells attach to food production surfaces. Live, damaged, or dead cells can attach to begin colonization. The conditioning layer starts as a thin, resistant layer of microorganisms, any combination of spoilage and pathogenic bacteria that form on and coat the conditioning layer [8]. As the layers of bacteria attach to the surface and to each other, they trap debris and nutrients, and the biofilm begins to take shape. Bacterial appendages (fimbriae, pili, and flagella) may also mediate attachment of other cells or materials to form the colony. *L. monocytogenes* likely attaches to surfaces by producing attachment fibrils [9]. During attachment, cells in the forming colony work together in a coordinated and cooperative function. This includes channeling nutrients to the film and removing waste products [2].

As the colony continues to attach, there is production of extracellular polysaccharides and changes in the cell morphology. Extracellular polysaccharide formation aids adhesion of the cells in the film and protects the bacterial layer against cleaners and sanitizers. The polysaccharide will also trap other cells and debris. Dr. Diebel states that the extracellular polysaccharide material form a bridge between bacteria and the conditioning layer with a combination of electrostatic and covalent bonds [1]. Development and growth, without removal intervention, results in the film becoming irreversibly attached to a substratum or interface or to each other, embedded in a matrix of extracellular substance that they have produced. Mature film reaches an equilibrium that delivers oxygen, food, and nutrients while carrying away fermentation products and sloughed cells. The outermost slime layer of film serves as a snare that traps additional contaminants and acts as a protectant, sealing the bacteria within so that the protected bacteria can be up to 100 times more resistant to sanitizer. As an example, *L. monocytogenes* in biofilms is more resistant to sanitizers than those not in films [5]. Inorganic and organic material flowing over the biofilm provide nutrient to the colony. Inside the biofilm, damaged or small cells may have the time to repair themselves and reproduce. The film is irreversible and now requires a special cleaning protocol for removal.

Biofilms form at a slow but steady rate and become harder to remove over time. They are most likely to form on rough penetrable surfaces, but they can form on just about any moist surface [8]. They can attach to all types of surfaces in food plants from stainless steel, especially on abraded or scratched surface, to polypropylene. Biofilms may form in hard-to-reach areas such as undersides of conveyor belts and seals. For this reason, it is necessary to regularly inspect and change equipment parts such as gaskets, o-rings, and piping. Where possible, food plants should identify and eliminate areas that cannot be thoroughly cleaned, through sanitary design of equipment. Extended production runs with minimal cleanup in between may increase the chances and frequency of films developing due to increased organic material contact time and formation of the conditioning layer. These longer runs also cut into valuable sanitation time and reduce the ability of sanitarians to do their job as designed and causing them to rush or take shortcuts. Additional causes may include the lack of stringent cleaning regime, pH extremes, and high contact surface temperature denaturing protein to facilitate formation of a conditioning layer, low fluid flow rate, and nutrient availability (Figure 5.1).

There are several problems, not the least of which is product contamination that occurs from the formation of biofilm. Product contamination occurs from

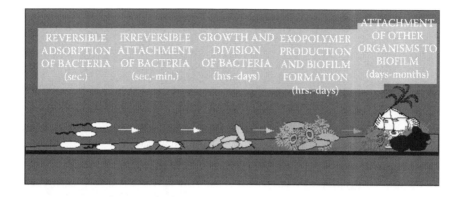

REVERSIBLE ADSORPTION OF BACTERIA (sec.)   IRREVERSIBLE ATTACHMENT OF BACTERIA (sec.-min.)   GROWTH AND DIVISION OF BACTERIA (hrs.-days)   EXOPOLYMER PRODUCTION AND BIOFILM FORMATION (hrs.-days)   ATTACHMENT OF OTHER ORGANISMS TO BIOFILM (days-months)

Figure 5.1   Biofilm displaying bacterial attachment magnified 10,000× USDA ARS. (From Arnold, Judy W., Bacterial Biofilms Less Likely on Electropolished Steel, *Agricultural Research*, 1998.)

sloughing bacteria that are shed periodically by the film and can reattach on equipment somewhere else in the product flow or make their way into food product. If these are spoilage organisms, product shelf life may be reduced and consumer purchases, especially repeat purchases, may decline. However, if they are pathogens, the product may be considered adulterated and subject to recall, or it may be responsible for a foodborne illness outbreak. Any company that has been involved in a recall or whose product has been associated with an illness can attest to the fact that they are damaging to the business and extremely expensive. Organisms within the film are also more heat resistant, so sloughed cells from films that form before cooking may not be destroyed as readily by the lethality process.

Other problems associated with formation of biofilms are accelerated deterioration of equipment through corrosion from cellular by-products. There may also be a reduction in the efficacy of heat transfer and impairment of detection devices as the film disrupts transmission [4]. As previously indicated, attached cells can develop increased resistance to cleaning chemicals and sanitizers possibly due to protection provided by the polysaccharide layer. The reduction in effectiveness of chemicals may be the result of cells layering and the reduction of exposed surfaces on which the chemicals contact. As an example, cells of *L. monocytogenes* in biofilms have been found to be more resistant to sanitizers than nonattached cells [10].

Evidence of Biofilm

There are several means of determining that a biofilm has begun to form on a food contact surface. Detection may be through the use of several senses. Visual signs include a "rainbow" appearance on stainless steel and tactile senses will detect a slimy feel on otherwise clean-appearing equipment surface. Although sour or off-odors may not indicate the presence of biofilms, they may indicate that a piece of equipment is not being cleaned thoroughly and that there is a potential for biofilm formation.

From an analytical standpoint, another indicator of biofilms is a sporadic spike in environmental test results due to bacterial sloughing. These may be found through generic microbiological tests such as APC or through environmental pathogen testing for plants conducting *Listeria* swabs. An increase in the bacterial counts or positive findings may indicate the formation of a biofilm and bacterial sloughing. Unfortunately, if they are not detected soon enough, the result may be sporadic product microfailures or decreased product shelf life. However, if you are already at a point where product has begun to fail shelf life or demonstrate higher than normal bacterial counts, it would be wise to consider the possibility of biofilm formation and apply control measures to eliminate them. ATP bioluminescence devices can be used to detect the presence of organic materials (Figures 5.2 and 5.3); unfortunately, ATP may not detect the presence of mature biofilms. The reason for this is that embedded cells do not move as much due to nutrient availability and use less energy, thus producing less ATP. Therefore, the device may provide a Pass reading on a surface where there is a biofilm.

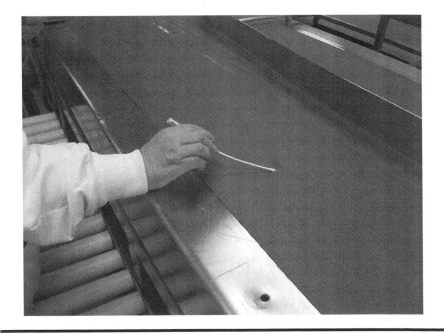

Figure 5.2   A clean line is swabbed using a premoistened swab.

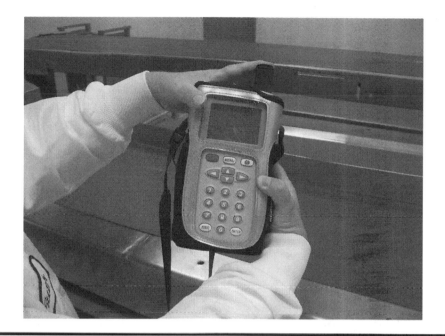

Figure 5.3   The swab is inserted into the ATP device to determine the presence of organic material that may indicate the presence of biofilms.

Biofilm Removal

There should be no good reason for the films to build if there is good sanitary equipment design and a regular and thorough cleaning to remove surface soil and subsurface film. Food manufacturing equipment poses many sanitary design challenges. Equipment has hollow rollers and tubing, welds, joints, and scrapers that make cleaning difficult. Once biofilms have established on a surface, they are harder to remove as they develop over time and require more aggressive action to eliminate. Fortunately, the original biofilm attachment is weak and easy to remove through proper sanitation procedures. Therefore, the film soil must be removed, and the most effective method of cleaning is the standard process described in Chapter 7.

1. **Dry clean.** This is done to remove as much visible soil or product material as possible. This may involve scraping, brushing, vacuuming, sweeping, or shoveling to remove large particles.
2. **Potable rinse.** The temperature of rinse water to remove initial soil should be between 120°F and 130°F to break down fats, but should not exceed 140°F to prevent creating baked-on soil conditions or mineral scale that will make removal of the biofilm even more difficult.

3. **Apply detergent.** In general, the use of a chlorinated alkali or a combination of oxidative agents and acids, such as hydrogen peroxide and paracetic acid, is recommended to break the chemical bonds of food soils. Depending on the food materials that the plant makes, the processes involved, and the soils created, it may be necessary to work closely with the sanitation chemical supplier to determine the specific combinations of cleaning chemicals and sanitizers to use. Application of the recommended chemicals over an extended exposure time (>5 min) will also be necessary to allow them to begin breaking down and removing the coating layer. Mechanical action (scrubbing on surfaces) or agitation, such as in a COP tank for small parts, is the most effective means of biofilm removal [8] and must be applied to completely remove the top layers of soil and subsurface attachment conditioning layer. Scrubbing with abrasive pads will help to remove breakdown the films for removal; however, the scrubbing must not be as intense as to etch or scratch equipment surface. Etching the surface only creates additional niches where films can form and makes removal more difficult.

4. **Final warm water rinse.** Again, the rinse water temperature should be <140°F to remove all cleaning chemical and bound soils.

5. **Sanitize.** Application of sanitizer should reduce any remaining bacterial count to negligible levels. Simply applying sanitizers to soiled surfaces is ineffective and a waste of money, since the efficacy of the sanitizer is reduced by the presence of soil. However, once the soil is removed and the biofilm is exposed, apply a higher concentration of sanitizer as lower levels will be less effective at killing the microorganisms in the film. As an example, apply sanitizer at 800–1000 ppm, and allow it to work for a period of time. Use a clear rinse and reapply sanitizer at the allowable level of 200 ppm without rinsing off. On weekends, apply quat at a level of 800–1000 ppm, and leave it on over the weekend to take advantage of its residual property. Acid-based sanitizers may be used to remove mineral film, and ozone use should also be considered as it is a strong oxidant that acts quickly against a wide array of microorganisms and has not been shown to result in organism resistance.

6. **Inspect.** As indicated previously, this will be through the use of physical senses, ATP bioluminescence, or microbiological testing to verify that the cleaning has been effective. Verification of Sanitation is covered in greater detail in Chapter 8.

Current Research

The USDA Agricultural Research Service (ARS) conducted studies concluded in 2006 that study biofilm formation and composition and included means of prevention and removal. ARS has already determined that a strong negative electrostatic charge to biofilms on stainless steel may reduce bacterial surface contamination. In the study, researchers at the Meat Quality Research Unit found that stainless steel surface-finishing treatments such as polishing, sandblasting, and grinding reduced buildup of biofilms [7]. They indicated that electro polishing, placing the stainless steel in an acid bath and running an electric current through the solution, prevented biofilm formation. The reason for this is that bacteria are negatively charged and the current through the acid media may change the charge on the metal, reducing the ability of bacteria to attach and form biofilms. A study by Dr. Michael Doyle at the University of Georgia Center for Food Safety, funded by the American Meat Institute, has shown that strains of lactic acid bacteria can inhibit growth of *Listeria* in a biofilm over extended time. The lactic acid bacteria did not grow at 39°F but did produce an anti-Listerial metabolite to keep levels of *Listeria* low. The future application of this information may lead to additional methods of both preventing and controlling biofilms. Until that time, food manufacturing plants will have to rely on methods that include sanitary design, effective sanitation, and monitoring for changes in the environment.

References

1. Deibel, Dr. Virginia and Jean Schoeni, Ph.D., Biofilms: Forming a Defense Strategy for the Food Plant, *Food Safety Magazine*, The Target Group, Glendale 2002/2003.
2. Swanson, Dr. Harley, *Understanding and Eliminating Biofilms in Food Processing Facilities*, National Meat Association Resource, Oakland, 2000.
3. Stier, Richard, Beating Back Biofilms in Food Processing, *Food Safety Magazine*, The Target Group, Glendale, 2005.
4. Stier, Richard, The Dirt on Biofilms, *Meat and Poultry Magazine*, 2002.
5. Molitz, Andrew G. and Scott E. Martin, Formation of Biofilms by *Listeria monocytogenes* under Various Growth Conditions, *Journal of Food Protection*, 2004.
6. Robbins, Justin B, Christopher W. Fisher, Andrew G. Moltz and Scott E. Martin, Elimination of *Listeria monocytogenes* Biofilms by Ozone, Chlorine and Hydrogen Peroxide, *Journal of Food Protection*, 2004.
7. Arnold, Judy W., Bacterial Biofilms Less Likely on Electropolished Steel, *Agricultural Research*, 1998.
8. Lupo, Lisa, Biofilms, *QA Magazine*, 2005.
9. Mustapha, A. and M. B. Liewen, Destruction of *Listeria monocytogenes* by Sodium Hypochlorite and Quaternary Ammonium Sanitizers, *Journal of Food Protection*, Vol. 52, No. 5, 1989, pp. 306–311.
10. Tompkin, R. B., Control of *Listeria monocytogenes* in the Food-Processing Environment, *Journal of Food Protection*, Vol. 65, No. 4, 2002, p. 720.

Chapter 6

Sanitary Facility Design

> I am not going to bankrupt the company spending money on sanitary design for a food safety hazard that is unlikely to occur anyway!
>
> **Anonymous plant manager, one month before the finding of *Listeria monocytogenes* in an RTE product.**

The concept of sanitary design is not entirely new to the food industry. Stainless steel was first developed in 1908 as one of the original steps geared toward sanitary design. Prior materials used for food manufacturing equipment such as iron, steel, and brass all corroded from the product, the process, or the environment. Many industry groups employed standards and pioneered improvements [11]. Since that time, many more improvements were made to food facilities and equipment for both efficiency and food safety. In the last several years, however, sanitary equipment design has taken on even greater significance in the industry as a means of providing greater assurance of safe food manufacturing.

As identified in Chapter 1, all food product manufacturers are required to comply with the FDA's Current Good Manufacturing Practices outlined in 21 CFR part 110.40, as well as state and local codes, to prevent the production of adulterated food products and ingredients. While there is not a clear linkage between sanitary design and the implementation of the Food Safety Modernization Act [23], there may be increased inspections by the FDA as well as more frequent control action if they find that food is being manufactured under insanitary conditions. In addition, facilities operating under USDA inspection must have in place Sanitation Standard Operating Procedures and meet Sanitation Performance Standards to ensure that food products and ingredients do not become adulterated during the manufacturing process, storage, or shipping. These standards are designed to be less prescriptive than in the past, allowing plants more flexibility as to how they achieve sanitation, including greater innovation in plant and equipment design. The industry also has experienced increased expectations for continuous improvement from retail, food service, and industrial customers as well as consumers. Thus, the burden for ensuring sanitary operations and preventing product contamination is where it belongs, with the manufacturer.

There are several reasons for the increased attention to using sanitary design for facilities and equipment to prevent microbiological contamination. First, consumers have shown an affinity toward minimally processed products and a preference toward Ready To Eat (RTE) foods for convenience. Second, there have been emerging pathogens of concern and an increasing population of consumers who are at higher risk when exposed to these organisms. Finally, there is a growth of large processors with nationwide distribution to meet these consumer needs, which can increase possible exposure of contamination to greater numbers of consumers [18]. Whether a company is building a new food manufacturing facility, expanding or upgrading an existing facility, or just maintaining their plant, sanitary design of the facility and equipment is one of the most effective food safety strategies. Incorporation of sanitary design into the facility can prevent development of microbiological niches, facilitate cleaning and sanitation, and the application of sanitary design can make for timely and effective cleaning of the company asset. The inclusion of design strategies that prevent the harborage or buildup of microorganisms or other food safety hazards may be more expensive up front. However, there may be benefits through reduction of time cleaning and maintaining equipment. In the long run, improved design may help maintain or increase product shelf life and improve product safety, reducing the potential of regulatory control action, foodborne illness or injury, product recall, lost revenue, or negative publicity.

Food manufacturers should start with the basic premise: food safety is nonnegotiable. Cleaning for food safety is the number one priority and can be achieved with good facility and equipment design. Food safety hazards controlled through sanitary design include microbiological (pathogens), physical (glass, metal shavings, wood), and chemical (allergen cross-contamination) while preventing product exposure to sources of filth (dust, rodent excrement). Prevention of food safety incidents means a company will not have to deal with negative publicity, but there is still a cost attached to hidden incidents. It is hard to show that food safety incidents have been successfully prevented, as it is hard to prove a negative. And it is very unlikely that food manufacturers will ever get the recognition in the media for the good things that they have done or continue to do to prevent foodborne illness, because preventing illness is not as interesting to the media as an illness story. Conversely, the consequences of food safety incidents are very well known: negative publicity, bad financial results, loss of customer and consumer confidence, and more restrictive regulations. While the temptation may be to try to cut expenses in the area of sanitation, never use hygiene or sanitation to balance the budget. Spend money up front on design rather than over time or to correct initial poor design.

In 2005, an AMI Facility Design Task Force, a multidisciplinary team from several food companies as well as design/construction firms, issued 11 Sanitary Design Principles for Facilities. The intent of these principles is to provide food manufacturers with guidelines for building and construction of food plants for the maximum prevention of microbiological contamination. Application of these principles is not limited to the United States. European Union (EU) legislation includes Machinery Directive 98/37/EC and Council Directive 93/43/EEC on hygiene of foodstuffs. The International Organization for Standardization (IOS) TC199 prepared "Hygiene Requirements for the Design of Machinery" that specifies hygienic standards for equipment design and performance. Finally, the European Hygienic Engineering and Design Group has developed design criteria for hygienic equipment and processing [17].

If a food company were to build a new facility, for cooked RTE products, it is recommended that they implement the elements of sanitary facility to ensure that they are able to operate under conditions that would be conductive to sanitary food production. Hygienic design principles can be applied to prevent contamination of food product from chemical hazards such as lubricants, coolants, and allergens and physical hazards by excluding foreign objects. While it may not be practical to incorporate all of these design principles into existing facilities, they can be included as facilities are remodeled, are upgraded, or as they expand. The principles developed by the AMI task forces are described below starting with the facility and continuing on to the equipment.

The AMI 11 Principles of Sanitary Facility Design [10,12]

Principle 1: Distinct Hygienic Zones Established in the Facility

> Maintain strict physical separations that reduce the likelihood of transfer of hazards from one area of the plant or from one process to another area of the plant or another process.

This means that the plant has conducted a hazard analysis to establish the needs for compartmentalization of processes in the plant and identified a logical process flow to protect product. Compartmentalization may involve separate locker and break room areas for employees in raw and RTE areas. Lockers themselves will be constructed of cleanable and durable material. The material should be able to withstand cleaning and sanitizing as well as application of pesticides. The tops will be sloped at a 60° angle to prevent dust accumulation or storage of personal or production items. The doors will be of a slotted or mesh design to provide airflow into the locker (Figure 6.1).

Figure 6.1 Example of plastic composition lockers, with vents and sloped tops to prevent accumulation of dust.

Restrooms should not open directly into production areas, and they should be provided with negative air pressure to avoid the spread of bacteria. Where possible, locate service areas such as parking lots, truck docks, and trash areas away from production areas to minimize potential contamination associated with these areas. Trash collection or compactor areas will be separate from ingredient or product handling and will have doors to prevent odors, insects, or airborne bacteria from entering the plant.

Principle 2: Personnel and Material Flows Controlled to Reduce Hazards

Establish traffic and process flows that control the movement of production workers, managers, visitors, QA staff, sanitation and maintenance personnel, products, ingredient, rework, and packaging materials to reduce food safety risks.

Environmental microbiological control in the plant is made more effective through separation of employees working in raw and cooked areas. This is accomplished by providing a separate entrance to the plant for employees and continues through isolation of lunchrooms, locker, and restrooms. Plant traffic patterns will be designed to prevent the entry of raw department employees and equipment from entering RTE areas. This may be accomplished by passive controls (cross-traffic aisles) or by active controls (card readers for RTE access) [12]. Plant tours are conducted in reverse order of the process; that is, they are conducted starting in RTE packaging and finish in the raw product area so that visitors do not transport contamination from raw to cooked. This also emphasizes to employees that raw and cooked separation is an expectation. It is desirable for the flow of the plant to be as straight as possible to prevent crossover from raw to cooked areas and to prevent cooked product from reentering raw areas. Smokehouses, for example, can be designed for product flow through, with raw product racks or trees entering from the raw side of the plant. When the cook process is completed, the racks or trees will exit from the opposite side so that they never return through a raw area. Batch kettle cooking is another process that can be set up much the same way with raw ingredients going into the kettle for cooking and mixing, and the cooked filling being pumped through a wall to a cooked product handling area.

Principle 3: Water Accumulation Controlled in the Facility

> Building systems, including floors, walls, ceilings, and supporting infrastructure, should prevent development and accumulation of water. Ensure that all water positively drains from the process area and that areas dry during their allotted time frames.

Water is used in many food manufacturing processes, and a significant amount of water is used for sanitation purposes. While it is a necessary component of food systems, it can also a necessity for bacteria and provides a mean by which bacteria can spread and contaminate product. Therefore, building design will prevent water accumulation by constructing effective floor drainage. The plant floor slope will be ⅛–¼ in./ft to drains approximately every 10 feet to prevent water accumulation. Drains will be pitched to flow from RTE areas to raw areas, and it is preferable that they do not connect but are separate systems. Do not use open trench drains, and eliminate poor drain repairs that prevent water from entering the drain. When possible, discharge water from drip pans directly to drains (Figure 6.2).

Principle 4: Room Temperature and Humidity Controlled

> Keeping process areas cold and dry will reduce the likelihood of growth of foodborne pathogens. Ensure that HVAC/refrigeration systems serving process areas maintain specified room temperatures. Control the air

Figure 6.2   Trench drains should be avoided in food plants as they are not easy to maintain and provide bacterial niches.

dew point to prevent condensation. Ensure that control systems include a cleanup purge cycle (heated air make-up and exhaust) to manage for during sanitation and to dry out the room after sanitation.

If the plant produces cooked RTE product, and it is a USDA-inspected facility, the processing areas are refrigerated to 50°F or less. A rule of thumb is that microbiological growth is reduced by 50% for every 10°F temperature drop. If the plant is maintained over 50°F, it may be necessary to conduct a cleanup after 10 hours or have a microbiological monitoring program in place to verify that microbiological growth is limited and will not result in high counts in the finished product. Provide sufficient filtered incoming air so that there is adequate makeup for any air exhausted by stacks or vents and control air temperature to prevent condensate. Condensation is the result of excess moisture that the air cannot hold beyond the saturation point, which is referred to as 100% relative humidity (RH) [21]. The excess moisture is just fog when it is in the air; however, collection on horizontal surfaces is condensation. Condensate forms when a surface temperature is lower than the dew point temperature of the air. If the material of the surface it forms on is somewhat porous, the moisture can penetrate the surface and become contaminated. Product or product contact surfaces may become contaminated when moisture condenses from air on contaminated surfaces and then drips onto product surfaces or product below [19].

Elimination of condensate is accomplished by warming the air above dew point or creating airflow that is sufficient to keep the air moisture limited.

Author's Note: The writer spent time consulting in a facility that was having some microbiological challenges with spoilage organisms. While evaluating the cleaning and sanitizing process one night, it was observed that the fog was so dense in the plant that the writer could not see from one line to another. This made it both dangerous for the sanitarians and made it particularly difficult for them to see the facility and equipment to determine whether they were doing an adequate job of cleaning. It also created a significant amount of condensation that had to be removed before the start of production, increasing the workload of the sanitation crew. Working with management and maintenance, the air handling units were operated in a way as to reduce the amount of fog in the plant and improve the safety and sanitary conditions for the sanitation crew.

Principle 5: Room Airflow and Room Air Quality Controlled

Design, install, and maintain HVAC/refrigeration systems serving process areas to ensure airflows from more clean to less clean areas. Adequately filter air to control contaminants and provide outdoor makeup air to maintain specified airflow. Minimize condensation on exposed surfaces and capture high concentrations of heat, moisture, and particulates at their source.

Air in the plant may come from several sources, depending on the type of process. Some facilities use fresh outside air through screened windows, ceiling vents, or hood vents. Others use mechanical devices for movement of air such as fans. The process of ventilation should remove smoke, steam, or odors from the plant, and bring in fresh, odor-free air. It is also used to provide a comfortable environment for plant employees [20]. In plants handling RTE product, air is a critical part of microbiological control. Airflow should be positive in RTE areas, meaning there should be greater pressure in the cleaner RTE rooms, from where the air should flow out to less clean raw or auxiliary areas. Incoming plant air is filtered at 95% efficient at 5 μm, which will remove bacteria [1]. There is sufficient makeup air brought into the plant to replace any air drawn out by fryer or freezer stacks or other vents. Compressed air will be filtered at 99.99% efficiency at 0.2 μm. It will be dehumidified to pressure dew point below the lowest ambient temperature to prevent moisture in the lines [15]. Use dust suppression equipment to limit airborne contaminants, especially allergens such as wheat or soy flour, and to provide added comfort for employees. Even with filtration, incoming air vents should be positioned not to blow air directly on product or product-handling equipment.

Principle 6: Site Elements Facilitate Sanitary Conditions

> Employ site elements that facilitate sanitary conditions. These include exterior grounds, lighting, grading, and water management systems as well as access to and from the site.

Sanitary design starts outside the facility, and the exterior of the facility should lend itself to sanitary conditions within the facility. This means that for all facilities, whether new or modified, the exterior and grounds can be maintained so that they won't contribute to contamination inside the facility. Parking and traffic areas should be paved to reduce the presence of dust. Avoid areas where contamination may come from adjacent areas such as sanitary landfill, refineries or chemical plants, salvage yards, and raw sewage treatment facilities. Grounds and yards should drain well to prevent pooling of water that can be an attractant for pests and breeding ground for mosquitoes. Lighting should be adequate to illuminate the facility but not attract pests. Placement of external lighting in locations away from the building will illuminate entrances but not attract insects to the building. This is presented in greater detail in Chapter 10. Where possible, use fences and gates, as well as swipe cards for plant door control for facility security.

When building a new facility, take into account the location, not the least of which is proximity to resources such as a good water source, labor, materials, and transportation. Availability of local emergency services, fuel, cleaning, uniform, and other services will also be part of the decision. The size of the structure and surrounding land are determined by near-term needs; however, it is may be wise to provide additional space for the expectation of future expansion. A facility that is locked in by adjacent property will be more difficult to expand than one that is surrounded by empty property or on a lot size sufficient for expansion. For the purposes of basic pest control, the grounds must not provide potential pest harborage areas. This is accomplished through landscaping design that provides an aesthetic appearance while preventing pest harborage and access. Grass and shrubbery are kept trimmed and away from the plant to supplement pest control.

Principle 7: Building Envelope Facilitates Sanitary Conditions

> All openings in the envelope, including doors, louvers, fans, and utility penetrations, should be designed to ensure that insects and rodents have no harborage around the building perimeter, access into the facility, or harborage therein. Envelope design and components should promote easy cleaning and inspection.

The building shell will be designed to prevent the entry of pests to the building and will be constructed to facilitate cleaning and ongoing inspection [10]. Prevention of pest access is covered in greater detail in Chapter 10; however, here are a couple of examples of prevention measures that can be included in sanitary design. Start with a facility foundation that goes down approximately 3 feet to prevent rats from gaining underground access [9]. Add a 30–36 in. wide and 4 in. deep strip of pebbles around the building to eliminate areas of pest harborage and hidden runway for rodents. This also provides easy access to perimeter bait stations for pest control operators.

Further exclusion of pests is accomplished through adequate door seals as well as screening of doors and windows with 22-mesh or finer material. Use of air curtains on doors that open to the outside or self-closing doors provides extra protection against pest entry (Figure 6.3).

Dock doors and dock seals will be secure to prevent insect and rodent access. The dock should have positive air pressure to prevent outside dust from entering the facility when the doors are opened for truck loading and unloading. Dock load lever plates will have brushes and seals to prevent pest access.

Roofing material is smooth and easy to clean and not coated with rock or pebble over tarpaper that can hold dust that can be pulled into the plant through air makeup systems. The preferred roofing material is single-membrane, pitched

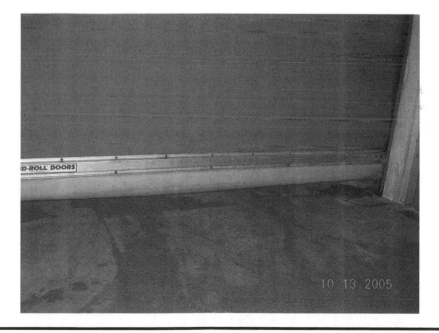

Figure 6.3 A self-closing door with a tight seal is good protection against the entry of pests and dust from the outside.

approximately 1 in. every 8 ft with evenly spaced drains to prevent accumulation of water [13].

Principle 8: Interior Spatial Design Promotes Sanitation

Interior space plan should promote the cleaning, sanitation, and maintenance of building components and processing equipment.

There should be sufficient space for cleaning around and under equipment and building elements. To minimize congestion, equipment should be no closer to the ceiling than 18 in. and 36 in. from walls or other equipment to and make cleaning easier. Gear boxes, motors, and drives are away from the product zone or, if over lines, shielded to prevent lubrication leaks onto product [7]. When motors are not on or near lines, they are placed on rails rather than flat platforms to prevent accumulation of soil and facilitate cleaning [15]. Where possible, group or centralize motor control centers and instrument panels in a single area away from production to reduce the number of rooms to clean [20].

Principle 9: Building Components and Construction Facilitate Sanitary Conditions

Design building components to prevent harborage points, ensuring sealed joints and absence of voids. Facilitate sanitation by using durable materials and isolating utilities with interstitial spaces and standoffs.

Materials used for original construction or renovation should be impervious, easily cleaned, and resistant to wear and corrosion [10]. Floors present a particular challenge to food manufacturing facilities. They are subject to extremes in temperature from freezers, fryers, hot water, harsh chemicals, equipment traffic, and foot traffic. Microorganisms on floors can get onto equipment from high-pressure water overspray during sanitation. Surface needs may differ in different areas of the plant; however, in the production areas there are several different options. The two most common are dairy tile or composite resin coating. Both are very durable and provide a cleanable, nonskid surface. Red-brick paver floors are laid with a waterproof membrane with a reinforcing fabric over concrete. Once the bricks are layered, the joints are sealed with epoxy grout [1]. These are very easy to clean, durable in wet and corrosive conditions, and hold up well in alternating hot/cold environment and under physical stress. Pavers are also easy to repair if a tile is damaged as the damaged tile can be removed and replaced with a new tile and resealed. Composite resin is also durable under similar conditions; however, over time the coating can crack and buckle. Some chemicals can quickly degrade epoxy and cold temperatures can cause cracks in floors, and once this happens, moisture pockets under the coating creating a microbiological niche (Figure 6.4).

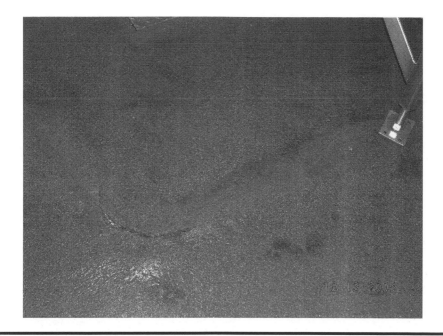

Figure 6.4   Cracks in flooring material can become bacterial niches and are difficult to maintain in sanitary condition.

Patching has limited impact as the water under the coating leeks to the patch and eventually results in the patch failing and other areas cracking. Wall–floor junctions should be coved with a 1–3 in. radius to eliminate a right-angled joint to prevent soil collection and facilitate cleaning (Figure 6.5). They must also be sealed with no cracks that can harbor food and moisture and create microbial niches.

If the junction is also curbed, the curbs will have a 30° slope to prevent accumulation of water, dust, or soil [18].

In addition to the materials for floors, walls and ceilings should be made of materials that are easily cleanable, nonporous, and resistant to chemicals and process conditions. The lower portions of food plant walls take a significant amount of abuse especially in work areas through the movement of pallets, vats, forklifts, and other equipment. For sanitary design and durability of the plant walls, there is a choice between insulated panels, Fiberglass Reinforced Panels (FRP), tile, concrete, or concrete block for the walls [18]. Here again, tile is a preferred material even though it is a little more costly. The benefit of tile is that it is durable, easy to clean, and simple to replace when damaged. If the plant chooses FRP or insulated panels, they will be as seamless as possible, but where seams are present, they will be caulk-sealed to prevent water niches. Walls will be solid from floor to ceiling, smooth, and nonabsorbent to prevent microbial or allergen niches. Wall-mounted control boxes and electrical panels will be mounted 1–3 in. away from the wall for cleaning access. If they were directly on the wall, they should be caulk sealed to

Figure 6.5 Spaces at wall–floor junctions must be eliminated or they can become harborage niches for microorganisms.

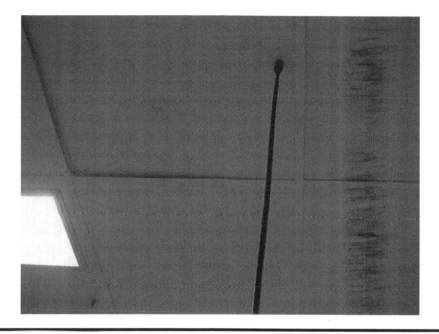

Figure 6.6 Track ceilings should be avoided unless panels are caulk sealed to prevent overhead soil from dropping onto product surfaces below.

prevent insect, soil, or microbial niches. Doorjambs at room entrances will be flush so that no ledges are created that will collect dust. Avoid installation of track ceilings with 2 × 4 foot drop-in panels. The tracks eventually flake and corrode, and the panels come loose from cleaning. This allows water from sanitation to collect above, provides area for dust accumulation, and ultimately becomes a microbiological and filth hazard. If track ceilings are already in place, caulk seal around the panels to prevent gaps and potential leakage of soils above the ceiling (Figure 6.6).

A preferred ceiling is a walk-on type with insulated glass board. This type of ceiling allows for creation of a loft for recessed lighting and utility runs so that hanging objects are minimized in production areas. It is best not to have penetrations for electrical or other utility drops; however, if they are required, seal at the penetrations to prevent leakage from the ceiling.

Principle 10: Utility Systems Designed to Prevent Contamination

> To prevent introduction of food safety hazards, utility systems should be constructed of materials that are cleanable to a microbiological level and provide access for cleaning, inspection, and maintenance. Systems should prevent water collection points, and prevent niches and harborage points.

Utility equipment should be designed using the AMI 10 Principles of Sanitary Equipment Design. All plumbing, airlines, ducting, and electrical conduit can be placed to travel through a structural ceiling loft. Do not route sewage lines to run over production or storage areas, rather have them routed through nonproduction areas so that leaks do not contaminate ingredients, product, or packaging. Make certain that the plumbing does not create any dead-end connections as material in the water can collect in the dead-end stubs creating the potential for contamination that may ultimately release and make its way into product. If dead-end pipe line stubs are present, they should be no longer than two times the diameter of the pipe to allow creation of turbulence to prevent settling in the dead end. All water lines for processing and sanitation will have backflow prevention devices to prevent potable water contamination. Air ducting or utility runs in the ceiling loft should be round, not flat, to minimize flat surfaces that can collect dust [2].

Principle 11: Sanitation Integrated into Facility Design

> Facilities should include integrated sanitation systems that control the introduction of hazards into the process areas such as hand sinks, hand sanitizers, doorway formers, boot washers, and foot baths, and those that facilitate cleaning and elimination of these hazards within the process area like hose stations, COP tanks, COP systems, and equipment washers.

Sanitation has got to be a consideration in the facility design to provide the right equipment in the right locations for effective sanitation. This means that there are a sufficient number of hose stations with hoses long enough to fully cover areas to be cleaned. A central cleaning system with automatic chemical dispensing is highly recommended. This provides accurate delivery of chemicals for effective cleaning and sanitizing as well as cost-effective use of materials. Clean in place (CIP) systems should be installed where possible (i.e., smokehouses, ammonia freezers), and Clean Out of Place (COP) tanks can be provided for small parts. Stainless racks are helpful to hold equipment parts so that no product contact parts or equipment is cleaned on or comes in contact with the floor. Provide storage areas for janitorial and sanitation equipment that have good airflow or are open to promote drying. It is important for sanitarians to have comfortable, dry gear to start their shift.

Other Considerations for Facility Sanitary Design

Hangers for Pipes and Conduit: There should be no exposed overhead pipes or conduit supported by angle iron, unistrut (Figure 6.7), or allthread as these materials create flat surfaces that can collect dust and plant soil. If angle iron is used, position it with the heel pointing up to prevent dust collection. It is best to use only smooth hanger rods and avoid the use of allthread.

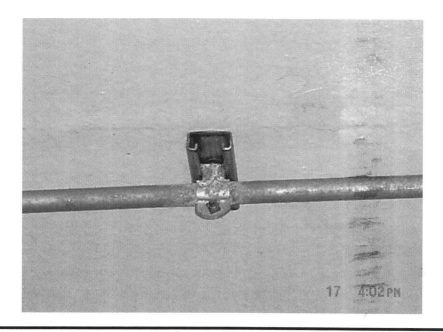

Figure 6.7   The use of materials like unistrut creates surfaces that can collect dust or soil and are difficult to clean.

Lighting: Where lighting is not recessed into a walk-on ceiling loft, the fixtures should be sloped and not flat as sloped surfaces will prevent dust collection. Lighting in production areas must be sufficient for cleaning, inspection, and for employees to effectively complete their work tasks. Though the needs may vary from production areas to storage areas, a general rule of thumb is light intensity of 25 foot 150 foot candles at 30 inches from the floor. Lights should be effectively shielded to prevent glass from contaminating product or product surfaces. The plant should have a "no glass" policy and a program for inspection and for handling glass breakage to prevent contamination.

Platforms and Mezzanines: Equipment or services placed on platforms and mezzanines require design features to facilitate cleaning and prevent contamination of surfaces below. The platform will be high enough off the floor to allow cleaning underneath. Platforms and mezzanines, including the steps or stairs used for access, should be on a solid surface, not an open tread. The surface typically selected is aluminum; however, the nature of the process must be considered so that the material selected will not rust. In wet operations, a checker or diamond surface will provide some measure of grip to prevent slips and falls. There must be a 4 inches curb height surrounding the surface to prevent kicking materials from the platform to areas below. The curb will be rolled from the plate material and covered to facilitate cleaning. They will not be raised from the platform as this will allow soil to fall to surfaces below. Where openings are present for equipment or connection lines, they will also have a 4 inch curb with a 4 inch gap to facilitate cleaning. If platforms are wet-cleaned, they should be self-draining but not onto equipment or contact areas. The stairs should be closed tread, have round handles to prevent soil contamination, and have risers without ledges that can trap soils [3]. Framing of the platform and stairs will use round square tubing turned 45° to create a diamond shape that will reduce dust and dirt collection. All tubing will be sealed at the end to prevent accumulation of food or water inside the tubing that can create a microbial niche [2].

Beams and Columns: If I-beams are used, they should have 60° closure plates to prevent dust accumulation. When they are used for columns, they can often be set too close to the walls for effective cleaning. Either move them 6 inches away from the wall for easier cleaning or fully enclose them to seal against the wall.

Equipment

In May 1996, FSIS proposed to amend federal meat and poultry regulations to eliminate requirements for FSIS prior approval of equipment and utensils used in official establishments. The final rule was published in the *Federal Register* in August 1997, thus removing requirements for prior approval of establishment drawings, specifications, and equipment. FSIS will continue to verify that establishments maintain equipment and utensils in a manner that will not lead to insanitary conditions and product adulteration. Under the final rule, FSIS indicated that manufacturers might want to consider using third-party certification services to ensure that equipment and utensils meet requirements for cleanability, durability, and inspectability. This process is voluntary, and the recognized services are identified in FSIS Notice 51-02, dated 11/26/2002 [6].

Prior to the preparation of the Sanitary Design Principles for Facilities, AMI had already involved a multicompany task force to prepare 10 Principles of Sanitary Design. This established recommendations for equipment to prevent harborage of microorganisms that can lead to product contamination or adulteration. Though the members of the task force are predominantly from meat and poultry manufacturing companies, there were also participants from food plant engineering groups and construction companies; thus, the principles can be utilized across a wide range of food manufacturing areas.

The 10 Principles of Sanitary Design

Principle 1: Cleanable to a Microbiological Level (Contact and Noncontact Surfaces)

> Food equipment must be constructed and be maintainable to ensure that the equipment can be effectively and efficiently cleaned and

sanitized over the lifetime of the equipment. The removal of all food materials is critical. This means preventing bacterial ingress, survival, growth, and reproduction. This includes product and nonproduct contact surfaces of the equipment.

Equipment should be designed so that surfaces lend themselves to cleaning and sanitizing. This means no horizontal ledges, recessed fasteners, or hidden areas [11]. Food contact surfaces will be free of breaks, seams, and cracks, and not impart color, odor, or taste. The ideal surface is smooth to be cleanable with no pores, free of crevices, sharp corners or angles, protrusions, or shadow zones [17]. It must also be hard, noncorrosive, and nontoxic and designed to protect product from foreign material [9]. An example of a surface that is difficult to clean and does not protect product from foreign material is a cloth conveyor belt. These are porous, making them difficult to clean and, because of the material, a constant source of fraying, increasing the potential for foreign material contamination in product. Interlock-type belts may be better for some applications provided belt hinges close on the conveyor belt bed to prevent accumulation of food particles and are open around sprockets to maximize cleaning. Hinge openings must be large enough to allow spray to reach bottom and top surfaces. The belts should be easy to remove for cleaning [14]. Some conveyor systems use UHMW caps on metal struts to minimize wear on the belt as it travels along the conveyor. These caps can become a niche contamination source and require removal for cleaning. Consider replacing the metal struts and caps with solid UHMW struts to eliminate the potential niches [16]. Eliminate dead ends or dead spaces, especially in pipes for food-conveying systems, as these becomes sources of accumulation [20]. Conveyor guides and splash guards are easy to remove for effective cleaning. Use of wood is to be avoided because wood is porous and cracks with age, making it nearly impossible to clean. Cracks in wood become a harborage for bacteria and chemicals. Because it can splinter as it dries and ages, wood also becomes a source of physical contaminants [3].

Principle 2: Made of Compatible Materials

Construction materials used for equipment must be completely compatible with the product, environment, cleaning and sanitizing chemicals, and the methods of cleaning and sanitation. Equipment materials of construction must be inert, corrosion resistant, nonporous, and nonabsorbent.

Food processing equipment must be able to hold up to the chemical composition of the product as well as cleaners and sanitizers used by the plant. The surface must be unaffected under the conditions of use in the plant [17]. A consideration for the plant is the product and associated risk, dry versus wet cleaning, and harshness of the process (i.e., brine tanks, extreme heat from fryers, or extreme cold from freezers). Do not use toxic materials for equipment such as lead, antimony, or cadmium [20]. Asbestos must never be used as this is a very hazardous material. Where it is found in older plants, it must be carefully sealed to prevent release into the atmosphere.

Principle 3: Accessible for Inspection, Maintenance, Cleaning, and Sanitation

Accessibility should be easily accomplished by an individual without tools. Disassembly and assembly should be facilitated by the equipment design to optimize sanitary conditions.

If an area on a piece of equipment cannot be seen or reached, it will be extremely difficult to clean or inspect for efficiency of cleaning. Equipment should be provided with quick release to minimize the need for tools to disassemble and to facilitate access to inner surfaces for repair, cleaning, inspection, and sanitizing.

Principle 4: Self-Draining—No Product or Liquid Collection

Equipment shall be self-draining to ensure that food product, water, or product liquid will not accumulate, pool, or condense on equipment or product zone areas.

Flat equipment surfaces should have a slight slope toward drain holes to prevent pooling of cleaners, sanitizers, or product moisture. If a slope is not possible, or if additional drain holes are not practical, there must be a procedure in place to facilitate draining and prevent pooling [17]. This may include wiping or using a squeegee on surfaces to direct pooled water off the surface or to a drain.

Principle 5: Hollow Areas Hermetically Sealed (No Penetration of Hollow Areas)

Hollow areas of equipment (e.g., frames, rollers) must be eliminated where possible or permanently sealed. Bolts, studs, mounting plates, junction boxes, nameplates, end caps, sleeves, and other such items must be continuously welded to the surface of equipment and not attached via drilled and tapped holes.

Hollow material is discouraged for conveyor rollers or equipment framing. If hollow material is used, it is to have a continuous weld seal (caulk is not acceptable). If penetrations are necessary, they should be weld sealed (Figure 6.8) at the

Figure 6.8 Penetrations such as bolts should have a continuous weld to create a seal to prevent harborage, not just a tack weld.

penetration and prevent food or liquid entry and creation of a possible bacterial or allergen niche.

Principle 6: No Niches (Pits, Cracks, Recesses, Poor Welds, Corrosion)

> All parts of the equipment shall be free of niches such as pits, cracks, corrosions, recesses, open seams, gaps, lap seams, protruding ledges, inside threads, bolt rivets, and dead ends. All welds must be continuous and fully penetrating.

All welds are continuous smooth and sanitary. Where seams are welded together, they are to be butt-welded, not overlapped, as this creates a void that can become a harborage niche. There should not be tack or spot welds that can harbor food or soil and become a microbial or allergen niche (Figure 6.9). Where spot welds are unavoidable, caulk between them for a continuous seal, though this is not as desirable as a continuous weld.

Welds will be ground and polished to the same texture as the surrounding surfaces with no globular welds as these can create bacterial niches. If polishing equipment is used for materials containing iron, they can transfer some of the iron

Figure 6.9 Tack welds should be avoided as they create areas that are difficult to clean. Welds should be continuous.

particles to stainless, eventually resulting in rust on the stainless material. It is recommended that at least one polisher be dedicated to stainless to prevent this cross-contamination with iron [15].

Principle 7: Sanitary Operational Performance (No Contribution to Insanitary Conditions During Operations)

> During normal operations, the equipment must perform so it does not contribute to unsanitary conditions or the harborage and growth of bacteria.

Avoid the use of screw threads and bolts in product areas, especially slotted, Phillips, and hex screws as these can trap food materials. Use polished stainless steel nuts to cover bolt threads [15]. It is recommended that the heads be sealed or that bolts come up from the underside, away from the product flow [17]. To prevent contamination that can be transferred from floor to equipment to product, equipment will be at least 6 inches above the floor. It will not be located near floor drains where aerosol or backups can present potential for equipment contamination. Where this is not possible, drains can be covered with stainless plates or rubber mats to prevent drain aerosols from backing up onto equipment surfaces.

Principle 8: Hygienic Design of Maintenance Enclosures (Junction Boxes, etc.)

Maintenance enclosures (e.g,. electrical control panels, chain guards, belt guards, gear enclosures, junction boxes, pneumatic/hydraulic enclosures) and human–machine interfaces (e.g., push buttons, valve handles, switches, touch screens) must be designed, constructed and maintained to ensure food product, water, or product liquid does not penetrate into or accumulate in or on the enclosure and interface. The physical design of the enclosures should be sloped or pitched to avoid use as a storage area.

Where possible, enclosures for water, air, or hydraulic lines should be designed into the equipment so that these lines are out of the product flow zones. Panel boxes should be designed to resist accumulation of water or soils internally, and have sloped tops so that they do not accumulate soil and cannot be used as platforms for materials that might make their way into product streams.

Principle 9: Hygienic Compatibility with Other Plant Systems (Electric, Air, Water)

Design of equipment must ensure hygienic compatibility with other equipment and systems (e.g., electrical, hydraulics, steam, air, water).

Connections are compatible, that is, of the same material, to prevent leaks or waste from equipment coming in contact with food-contact surfaces. Use of dissimilar materials at connections can result in electrolytic corrosion from use or from chemicals and should be avoided [20].

Principle 10: Validate Cleaning and Sanitary Protocols (Encourage Equipment Designers to Demonstrate Effective Cleaning of the Equipment)

The procedures prescribed for cleaning and sanitation must be clearly written, designed, and proved to be effective and efficient. Chemicals recommended for cleaning and sanitation must be compatible with the equipment, as well as compatible with the manufacturing environment.

Prior to installing any equipment, make it a point to work with equipment vendors to ensure sanitary design and to have the design reviewed by a third-party expert. Experience has taught the industry not to assume that equipment manufacturers are going to do this. A cross-functional team (QA, Sanitation, Maintenance,

Production) should be used to evaluate pieces of equipment before they are purchased and to develop very specific sanitary requirements for all plant equipment. These are the individuals charged with the cleaning, inspection, operation, and maintenance of equipment, and they should be part of the decision-making process in choosing equipment.

Other Considerations for Sanitary Equipment Design

Bearings and Shafts: Moving parts such as drive shafts should be sealed with self-lubricating bearings that do not leak. Lubricants should not be applied so heavily that they leak onto product surfaces [20].

Equipment Legs and Frames: If the frames have rolled edges, they should not exceed 180° to prevent areas that can collect moisture and soil. If they already exceed this tolerance, close them all the way and weld seal them to prevent niche sources. Equipment legs may need to be adjustable to facilitate leveling; however, they should have minimal penetrations. The base should not be a source of contamination at the floor junction and may require sealing or the use of cone covers to prevent irregular niche areas (Figure 6.10).

Figure 6.10 Equipment bases such as this will be difficult to clean and may create a bacterial harborage niche.

Repairs

Repairs in food plants should be permanent and sanitary. Temporary repairs should be avoided whenever possible. Temporary repairs are just that—temporary. Unfortunately, they have a way of becoming permanent if they have no follow-up. It is understood that temporary repairs may need to be made on occasion; however, the plant must have a system to track the date of the temporary repair and a means to follow up with a permanent repair. For example, a minor leak from the roof or an overhead pipe might be repaired by hanging plastic sheets to catch water from the leak or to direct it away from production areas and to drains. The date the plastic is hung should be written on the plastic with a marker pen and must be replaced each day until a permanent repair can be implemented. Under no circumstances should repairs be made with duct or plastic tape as this becomes a very trap for soil and microorganisms. Avoid the use of temporary repairs with paper- or plastic-coated twist ties as these are difficult to clean and become a potential physical product hazard.

Caulk has limited application for either permanent construction or temporary repairs. It can be functional when sealing along fiber wall paneling to prevent water from getting in between seams. It may also be used around ceiling panels used in a track system as a means of preventing dust and soil that collects above the panels from dropping down onto the product zones below. However, keep in mind that it may eventually dry and shrink, resulting in it becoming loose and will need to be replaced over time. When using caulk, it must be compatible with the substrate and the surface must be clean and dry when it is applied [6] (Figure 6.11).

In Europe, there is a trend toward creating production plants where each processing area is an entity unto itself with positive air pressure and independent temperature and humidity control. There are several other standards that European food manufacturing plants employ that could be considered in the United States. For example, European plants are using cloth air ducting as opposed to galvanized material. The cloth ducts are able to be removed periodically for cleaning, and this reduces the buildup of dust or other solid on the overhead structures. European requirements for drains are one every 3 m² as opposed to one every 10 feet. This results in more complete drainage of floor areas. Finally, there is a preference for white walls and floors that creates an environment where cleanliness is very visible [22].

Careful planning can provide greater assurance of food safety. Cross-functional training of staff in sanitary facility and equipment design can enhance evaluation of existing structure and plant equipment or to facilitate expansion and improvements. This can be accomplished through use of available literature, or more effectively, through training courses offered by experts in the field. Use a cross-functional design team comprising Production, Quality Assurance, Maintenance, and Sanitation to establish requirements of sanitary design with the desired outcome being a facility and equipment that supports food safety, quality, and productivity.

There are many other benefits to effective design that can offset expenditures for sanitary materials and equipment. Materials used in facility design can be durable,

Figure 6.11 Incomplete application of caulk is insanitary, creates bacterial potential, and is not aesthetic.

reducing repair and maintenance costs. The same holds true for equipment design. Both can have an impact on sanitation expense by reducing the amount of chemical needed to clean and sanitize, as well as minimizing labor and overtime. Sanitary design makes cleaning faster and more efficient. The easier the facility and equipment are to clean, the better the job will be done. However, the most important factor, product protection, is the most compelling reason for following basic sanitary design principles. In the long run, this will be a benefit to the company, so make sanitary design part of the overall plant food safety system.

References

1. Graham, Donald J., Using Sanitary Design to Avoid HACCP Hazards and Allergen Contamination, *Food Safety Magazine*, The Target Group, Glendale, 2004.
2. Graham, Donald J., *Sanitary Design and Construction Course,* Graham Sanitary Design Consulting, Ltd, 1999
3. Imholte, Thomas J. and Tammy Imholte Tauscher, *Engineering for Food Safety and Sanitation: A Guide to Sanitary Design of Food Plants and Food Plant Equipment,* Technical Institute of Food Safety, 1999
4. U.S. Department of Agriculture, *Sanitation Requirements for Official Meat and Poultry Establishments*, Washington, D.C.

5. U.S. Department of Agriculture, FSIS Directive 5000.1, *Sanitation Performance Standards*, Washington, D.C., 2000.
6. U.S. Department of Agriculture, FSIS Notice 51-02, *Third-Party Equipment and Utensil Inspection and Certification Programs for the Meat and Poultry Processing Industry*, Washington, D.C., 2002.
7. Gabis, Dr. Damien A., *Hygienic Aspects of Food Processing Equipment*, Scope, Silliker Laboratories, 1996.
8. Troller, John A., *Sanitation in Food Processing*, Academic Press, 1983.
9. Gould, Dr. Wilbur A., *CGMP's/Food Plant Sanitation*, CTI Publishers. 1994.
10. Seward, Dr. Skip, How to Build a Food-Safe Plant, *Meat Processing*, April 2004, p. 22.
11. Stauffer, John E., *Quality Assurance of Food Ingredients Processing and Distribution*, Food and Nutrition Press, 1994.
12. Bricher, Julie Larson, A Blueprint for Success in Sanitary Equipment and Facility Design, *Food Safety Magazine*, The Target Group, Glendale, 2004/2005.
13. Pehanich, Mike, Blueprint for Food Safety, *Food Processing*, March 2005, pp. 51–58.
14. Anon, Selecting Easy-to-Clean Conveyor Belts for Superior Sanitation, *Food Safety Magazine*, The Target Group, Glendale, 2005, p. 46.
15. Graham, Donald, *Equipped for Excellence, Sanitary Facility Design Essentials*, presented at Food Processing Machinery Convention, Las Vegas, 2005.
16. Rosen, Joan, *Sanitary Design Principles in the Fresh-Cut Produce Industry*, presented at Food Processing Machinery Convention, Las Vegas, 2005.
17. Curicl, Roy, Building the Self Cleaning Food Plant, Hygienic Design of Equipment in Food Processing, *Food Safety Magazine*, The Target Group, Glendale, 2003, pp. 51–53.
18. Graham, Don, *Sanitation Design*, presented at Foodborne Pathogen Briefing, Chicago, January 7, 1999.
19. Anon, *Condensation: Solving the Problem*, Hendon and Redmond, website resource, 2005.
20. Katsuyama, Allen M., *Principles of Food Plant Sanitation*, Food Processors Institute, New York, 1993, pp. 205–208.
21. Anon, *Condensate and Dripping Water*, National Meat Association Resource, Oakland, 1996, p. 1.
22. Higgins, Kevin T., Standardized Sanitation, *Food Engineering*, August 2012, pp. 59–68.
23. Lupo, Lisa, Sanitary Design of Equipment, *Quality Assurance Magazine*, March–April 2012, pp. 44–47.

Chapter 7

Sanitation Best Practices

Since QA is making us clean our tools, they should pay for the chemicals!

**Anonymous plant maintenance employee upon
finding out he has to clean and sanitize his tools.**

Why clean the plant? The time used in cleaning a food plant is time away from production, and it is production that makes money for the company, correct? Production and sale of finished product does make the company money, provided the production is wholesome, unadulterated, and of a quality level that people will continue to buy it. Without a sanitation process in a food plant, it is likely that none of these expectations will be met. Sanitation is basic to food safety and quality, and it is a vital segment of an integrated food safety system with strong links to regulatory compliance, quality, HACCP, GMPs, and pest control.

The process of sanitation has many facets that make it vital in a food plant, not the least of which is that it allows food companies to meet regulatory standards. Of course, the primary function is to remove contaminating soils, prevent film buildup, and prepare the food surface for sanitizing. It is also necessary to prevent insect and rodent infestation and harborage by removing sources of attraction and nutrition. Effective sanitation also plays an important role in preventing allergen cross-contamination and foreign material inclusions. The benefits of effective sanitation are production of safe product, improved product shelf life, and reduction of off-flavor, odor, and color. To an extent, it will also prevent equipment deterioration and increase production efficiency. Finally, it can be a source of pride and morale to employees who prefer to work in a location that is clean.

Food plants operate under federal regulations regarding the maintenance of sanitary processing conditions. Facilities that do not maintain those conditions may have their products retained, their production delayed, or their operations suspended. It is essential to remove soil and microorganisms during the sanitation process to prevent the manufacture of foods under insanitary conditions, rendering it unfit or injurious to health and leaving the company open to regulatory enforcement action. The cost of poor sanitation is product on quality hold, increased storage costs, customer order shortage, production stoppage, material disposal, customer/consumer complaints, potential liability and associated costs, and possible product recall or seizure. Facilities must operate under sanitary conditions and in a manner that ensures product will not be adulterated. For most operations, this means that the plant equipment and environment shall be cleaned at least daily (within a 24-hour period) or as frequently as needed to prevent insanitary conditions or product adulteration.

Product safety and quality is highly dependent on sanitation, as improper sanitation will result in reduced shelf life and increased loss due to spoilage. Sanitation is a prerequisite to HACCP and is intended to reduce incidence of microbiological, chemical, and physical hazards in the food manufacturing environment. The most effective sanitation program can be nullified if employees do not follow good manufacturing practices and create contamination conditions. Conversely, strong sanitation programs, incorporating multiple interventions, integrated with other critical food safety systems will enhance overall product safety. This chapter will introduce the basic needs for effective sanitation and the need for a strong linkage to other food safety programs in the plant. It will touch on the benefits and incentives for a food company to implement robust sanitation practices and link them with the total food safety system. This chapter will provide basic operational guidelines pertaining to sanitation practices and sanitation employee expectations for properly cleaning a food plant: its equipment, utensils, and structure.

Who Is Responsible for Sanitation?

While the primary focus is typically on the sanitation department, effective sanitation involves a combination of efforts by multiple departments. Sanitation actually begins with the implementation of sanitary design of the plant and manufacturing equipment, and sanitary design involves cross-functional efforts between engineering, maintenance, sanitation, quality, and operations. Sanitation continues with the development of an effective sanitation plan, and then the implementation of that plan, which also involves several of the plant or company departments. However, the plant manager is ultimately responsible for the implementation and enforcement of the sanitation requirements. He or she controls the budget for sanitation supplies, training, and equipment. They usually initiate the process for capital expenditures for new sanitation equipment or physical plant improvements that can make sanitation more effective. Top-down management support is vital to set the tone for the perception of sanitation priority within the plant, and if they demonstrate a commitment to sanitation and the sanitation crew, their direct reports will generally recognize sanitation as a priority as well. They also should take responsibility to ensure that the entire plant understands their role and responsibility to sanitation and product safety.

Quality assurance (QA) provides guidance in the development of sanitation procedures and will verify that procedures are followed and that the manufacturing environment is clean. QA must be provided the authority to reject equipment, lines, or rooms deemed not meeting sanitary requirements, without reprisal or intimidation. QA is usually responsible for collecting microbiological swabs to verify the cleanliness of equipment; thus, they have access to data regarding sanitation performance. This data should be shared with plant management and sanitation management to let them know where performance is satisfactory and where improvement is needed. This will be covered in greater detail in Chapter 8. The sanitation manager or supervisor is responsible for training the sanitarians, for implementing the requirements of the sanitation operation procedure and the master sanitation schedule, and for verifying sanitation efficacy.

Sanitarians

The sanitation department is one of the most important departments in a food manufacturing facility. How can you start your plant in the morning, on time, without it being clean? How can you expect to have acceptable product quality and shelf life without a sanitary manufacturing environment? And, how are you going to be certain that your products are safe for consumption without an environment nearly free of pathogens? You cannot do it without the actions of a group of individuals willing to work in the dead of night, often in wet conditions, usually with little to no recognition unless something goes wrong.

The sanitation department depends on structure to be effective. This begins with a strong sanitation supervisor or manager; someone with good leadership attributes, technical skills, and problem-solving ability. It continues with the selection, hiring, and training of sanitarians that implement the sanitation process. How do you make sure that we have put sanitation in a position to be the most important department in the plant? Start by hiring a staff that is physically able to do the job, that can work the hours required and can handle the conditions (wet, chemicals). Interview potential sanitation department candidates just as you would candidates for any other position. When interviewing the perspective candidates, look for associates who will be conscientious, take pride in their work, and be safety conscious. They must also have the ability to work with chemicals and mixing. It may also be necessary to hire sufficient sanitarians to account for vacations or illnesses as well as people who decide not to work on this shift [13]. The sanitation job is challenging enough without being short-staffed. As with any other position in the plant, it is also a good idea to cross-train sanitarians to ensure that they can function in different roles when short-staffed.

Sanitarians must be well trained to do their jobs, including how to perform their tasks safely. Appropriate personal protective equipment must be provided for and used by associates, such as rain gear, aprons, boots, gloves, and goggles or face shields. Inappropriate garments should not be allowed (i.e., street clothes, garbage bag coverings, or flammable material) as these can result in poor safety or sanitation conditions. Do not just give sanitarians the personal protective gear without telling them why it is necessary for them to wear it; train them to understand why it is important to wear this gear. In one plant, the writer observed that plant sanitarians wore street clothes, had no gloves or goggles, and used trash bags to fashion covers to keep themselves dry. When asked why they did not have appropriate rain gear, the plant manager explained that it was too expensive. A review of the plant records revealed a high frequency of reportable accidents from chemical burns and irritations among the sanitation staff resulting in a high insurance premium, and lost time among the sanitation staff resulting in late plant start-up. It did not take long to calculate that providing the appropriate gear was less expensive than the dollars spent on insurance and lost productivity. The plant manager relented and bought the appropriate equipment, but it was unfortunate that people had to be injured before he took action. In another plant, the manager used written procedures and pictures to teach sanitarians about lock-out and tag-out procedures, what they are, and why they are important. He made sure that they knew about and had access to material safety data sheets (MSDSs) and that they had access to chemical spill kits. Not surprisingly, this plant not only has had minimal injuries and expenses for the sanitation staff, but they also start on time and have one of the lowest frequency of microbiological failures, both environmental and product.

Lock Out Tag Out (LOTO) is a process that is extremely critical to the safety of plant sanitation employees or other employees who may be working on equipment where hazardous energy could result in serious injury or death. The program is designed to control the energy to prevent unexpected release or accidental start-up of the equipment. There are three basic steps to implementing a successful LOTO process, and they are identified as follows:

1. Identify the risks associated with the equipment that is to be cleaned. The risks must include the types of energy such as electrical, hydraulic, gravity, springs, etc. Once the risk has been identified, then the determination must be made how the risk can be minimized or eliminated through control of the energy. This may be accomplished through actions as simple as turning a switch or a breaker to the "off" position to more complicated steps such as bleeding off the energy or blocking the source.
2. Once the risks and the controls have been identified, then detailed written procedures must be developed. Each procedure must identify the specific piece of equipment to be controlled and step-by-step directions for the LOTO

process. It is not required that pictures be included; however, for the clarity of the procedure, it is highly encouraged. The specific method or location of the control device must be identified. If energy must be dissipated, this must also be clearly identified in the procedure. Once the procedure has been completed, it is necessary to test the process to ensure that it works. This is accomplished by going to the specific piece of equipment, implementing the LOTO process, and then testing to ensure it has been successful by re-energizing, turning the equipment ON, and ensuring that the equipment does not operate. Once completed, then turn the equipment to the OFF position. If the procedure does not work, then it must be revised to ensure that it is effective.

3. Finally, once effective procedures have been completed, the employees who will be implementing the procedure must be trained. LOTO training, like any training performed, must be documented for assurance that each employee has received the appropriate training and to verify that they comprehend the material. In addition, it is advisable to evaluate employee implementation of the procedures as an additional step to verify comprehension. If it is noted that the employee doesn't fully understand the steps, then there is an opportunity to conduct retraining. If on the other hand the procedure is not effective in removing the energy hazard, there is another opportunity to modify the procedure.

The consequences of failure to implement effective LOTO procedures could be injury or fatality of a valued employee. Additional consequences include citation and fines by OSHA if an investigation reveal missing or inadequate procedures [20] (Figure 7.1).

The sanitation process is defined for sanitarians through concise sanitation operating procedures. It is vital to provide sanitarians with clear, written procedures for each area and piece of equipment to be cleaned so that they can understand plant expectations of cleanliness. This manual is a valuable tool for training of sanitation personnel. It can be very effective to incorporate photos to show proper technique. If your company has an intranet, digital photos and even video can be incorporated into the online procedures for even more effect. The plant sanitation supervisor should be the person to conduct training session for new employees as well as provide ongoing (updated) training to existing crew. The chemical supplier's representative and QA manager can also assist in training sanitation crew members. When writing sanitation procedures, involve personnel from the sanitation crew since they are the most familiar with the process and equipment. This gives them a sense of ownership and helps to standardize the processes. Use this information to ensure that you have provided your sanitation crew with all that they need to continue to be the most important department in the plant. The same expectations are true for the members of the day sanitation crew. Although they may not use cleaning chemicals, they are often responsible for knowing how to monitor and change sanitizer in hand dips or shoe dips. However, keep in mind that training is

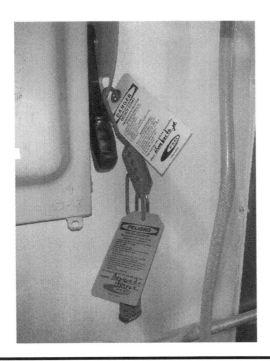

Figure 7.1   This demonstrates the application of a lock to the energy source of a piece of equipment prior to the cleaning of the equipment.

at many levels and involves more than a lecture, plugging in a video, or providing some reading material. Training encompasses all senses: hearing and visual, hands on, feedback, and follow-up. Training records for each person must be up to date, signed by the associate, and readily available for audit at all times. These will be maintained by plant human resources or the sanitation supervisor.

Some plants choose to use an outside contract cleaning and sanitizing company to perform sanitation activities. There are certain advantages to using these resources as they are responsible for providing the crew, which can be helpful when local resources are scarce. There may be a financial benefit to outsourcing that the company will have to explore. The contract crew can be used to help maintain the Sanitation Standard Operating Procedures (SSOPs); however, the company is ultimately responsible for compliance with all sanitation requirements and regulations.

The sanitation process is enhanced through the use of sanitary technique to effectively clean the food plant environment as well as housekeeping in auxiliary storage and office areas. When possible, separate sanitarians so that part of the crew is dedicated to cleaning raw areas while others are dedicated to cleaning RTE areas. If it is not practical to have separate crews, then make certain that there is a decontamination process in place if they move from cleaning a raw area to cleaning an RTE area. This may include changing their sanitation gear, changing or cleaning and sanitizing boots, and washing their hands. Make sure that they are aware of and comply with the GMPs identified in Chapter 9. Provide information on integrated pest control processes and the sanitation department role in pest control, as identified in Chapter 10, as it is often under the responsibility of the sanitation department and is closely linked to sanitation efforts.

Provide sanitarians with an environment and equipment that will be relatively easy to clean (sanitary equipment design, solid floor surfaces, no spot welds, etc.) in the time allowed. Steam fog due to inadequate exhaust is not conducive to cleaning and inspection as well as being a potential safety hazard, so make sure that the facility has adequate ventilation for the sanitation process. Provide the tools to effectively clean and sanitize and give them the training and direction on how to use these tools to do the job properly and do not spare on equipment or chemicals. Approximately, $0.80 of each dollar spent on sanitation is spent on sanitation labor [9], so invest in sanitarians and do not cut sanitation expenses to balance the plant budget. Give them enough time to clean the environment you have provided. In other words, make sanitation easy or the sanitarians will by taking shortcuts that may have a negative impact on sanitation and, ultimately, on product. Give them opportunity to meet with the sanitation chemical representative and let them discuss technique, the best chemicals for the job, and the needs for equipment. Let the sanitarians help make decisions on the best equipment for the job that they are doing (e.g., belt washers, Clean Out of Place [COP], or Clean in Place [CIP] systems).

Provide the sanitation department with the right people, the right tools, and the right training and you are well on your way to an effective program. Also provide them with the time to do their jobs. Many companies are under pressure to increase throughput and reduce cost. Most often, the means by which runtime is increased is by reducing cleaning time. While it is important for a company to increase operational efficiency, this may decrease the time allotted for sanitation. In these instances, the change in the production/sanitation ratio may result in ineffective cleaning and sanitizing, increased product failure, potential foodborne illness liability, and possible regulatory control action. If there is an increase in the production day, review this change with the sanitation personnel and determine what changes they may need to do their job effectively. This may include an increase in the sanitation staff, use of automated cleaning systems, or redesign of equipment to facilitate effective cleaning.

By all means, do not forget to give the sanitation department recognition too. They often miss out on the parties and the company lunches and do not see management or hear about the status of their company. Give them feedback on how they are performing (timely start-up, microbiological levels, product shelf life, etc.) to give them a sense of contribution. It is also valuable to spend some quality time on the night shift, provide pizza or cater holiday meals, celebrate monthly birthdays, review microbiological swab data with them, let them know how their company is performing, and remind sanitation associates how important they are and that you support their efforts. This applies to the plant manager as well as the QA manager. Remember, if they are successful, your business will be successful.

Written Cleaning Procedures

Before preparing the sanitation manual, it is best to start by making sure that all plant management, including production, maintenance, QA, and sanitation, understand the basics of food plant sanitation. Sanitation is essential to food safety and quality and is the foundation for all food safety programs. Sanitation improves operational efficiency, productivity, and quality, as well as promotes employee safety. It can be an indication of how well an operation is run. The objective is to remove soil and microorganisms from manufacturing environment to prevent manufacture of foods under unsanitary conditions. If done correctly, the result is the destruction of microorganisms, including pathogens, spoilage, and indicators. When this is accomplished, you are more likely to (1) prevent foodborne illness, (2) reduce product spoilage, and (3) increase product shelf life. Each of these elements should contribute to the company reputation and earnings.

Each plant shall have a complete written sanitation manual to cover all cleaning processes, whether by hand, COP, CIP, or other method. This manual also shall cover cleaning of the entire plant environment in addition to processing equipment. It likely will be more detailed than SSOPs and can be used as a training manual for sanitarians as well as for plant employees. While it is best that the manual be in English, it may be necessary to provide copies in other languages represented by the employees in your plant. These procedures must be accurate and realistic, given the nature of the operation. The manual will define the procedures to be followed to destroy vegetative organisms of public health significance (pathogens), reduce the numbers of undesirable organisms (spoilage, indicators), and eliminate the conditions that allow their growth. It should begin with basic safety practices for sanitation employees, including

- The use of personal protective equipment, when and how to wear them.
- Lock-Out/Tag-Out procedures.
- Basic plant safety precautions (no running, safe lifting procedures, etc.).
- Chemical safety procedures (proper mixing and application, proper storage, proper container labeling, etc.).
- Basic first aid for chemical exposure.
- It would also be valuable to include procedures for chemical spill control.

Procedures will include a complete layout of the plant and equipment, map of drain location (drains being numbered) and common drains, and the responsible personnel for each area. It will also identify the locations of hose stations or chemical hookups. It will include a listing of chemicals, their usage levels, a copy of their labels,

Table 7.1   Master Sanitation Schedule

Cleo's Foods Master Sanitation Program				
Task Description	*Frequency*	*Date Scheduled*	*Date Completed*	*Sanitor*
Clean overhead pipes and beams	Weekly			
Clean locker rooms	Monthly			
Rod and flush drains	Quarterly			
Clean electrical panels	Weekly			
Clean refrigeration units	Weekly			
Clean ceilings and overheads	Weekly			
Remove and soak conveyor belts	Monthly			
Scrape ice in freezers	Monthly			
Empty and clean coolers	Monthly			
Empty and clean outside storage tanks	Semi-annually			
Scrub warehouse floor	Quarterly			

and MSDS sheets. The manual will provide detailed cleaning instructions for each piece of equipment: the need for disassembly, the handling of parts in a COP tub or on a cleaning rack, the processes for CIP systems, and the processes by which cleaning is accomplished. The use of diagrams or photographic materials can be used to facilitate description, provided the photos are treated as confidential. The manual may also include non-daily cleaning procedures that are part of the master sanitation schedule (Table 7.1). A master sanitation schedule lists daily, weekly, monthly, quarterly, etc., cleaning tasks and is prepared and maintained by the sanitation department. It will show the cleaning task, scheduled cleaning date, completion date, and sanitarian signoff to indicate that the procedure was implemented. Procedures will include instructions for the cleaning and storage of sanitation equipment.

Considerations for Effective Cleaning

When developing sanitation procedures, as when selecting cleaning compounds, there are several considerations that must be made. First is the type of soil to be

Table 7.2   Food Soil Solubility and Recommended Cleaner

Food Soil	Solubility	Impact of Heat on Solubility	Recommended Cleaning Compound
Carbohydrates	Soluble in water, usually easy to remove.	Heat caramelizes carbohydrates and makes them more difficult to remove.	Alkali.
Proteins	Less soluble in water when undenatured.	Heat denatures protein and makes it more difficult to remove.	Alkali. Acid may be used.
Fats	Least soluble in water. Breakdown with heat.	High heat makes fats more difficult to remove.	Alkali. Emulsified by phosphates.
Minerals	Insoluble in water and alkali.	Heat and water hardness increase difficulty in removal.	Acid removes mineral films (milk stone and calcium oxalate).

cleaned, the second is the function of various chemicals in the cleaning process, and finally the condition of the plant water.

Soils

There are many types of soils that may be encountered in food plants, depending on the types of product being made. Each different soil has a different level of solubility. There is not an all-purpose cleaner to address all soils; no "one size fits all." Cleaners or detergents are selected specific to the needs [2]. Table 7.2 illustrates the solubility of various soils typically found in food processing plants [2–4].

Chemical Functions

Other important factors to consider when setting up the sanitation program is the functionality of chemicals; that is, what they do when used. The following are functions that cleaning chemicals perform when used in a sanitation system [2].

■ *Emulsification:* This is the breaking up of fats and oils to allow them to mix in water. Once this is accomplished, they remain suspended in water until rinsed away.

■ *Saponification:* This is the process of making fat soluble and easier to remove. Alkali cleaners react with animal or plant fat creating soap that is suspended for rinsing.

■ *Sequestering/chelating:* This is the process of removing mineral hardness from water and making water softer for cleaning. Polyphosphates are examples of sequestering/chelating agents.

■ *Wetting agents:* These are used to lower the surface tension of water, helping the water to contact all surfaces of the soil and the equipment.

■ *Penetration:* This is needed for effective soil wetting and occurs as liquids enter materials through pores and channels in porous material.

■ *Dissolving:* This chemical reaction produces water-soluble product from water-soluble soils. Some soils, such as alkali deposits, form strong bonds with surfaces. Acid will solubilize these soils for removal.

■ *Dispersion:* Also known as *deflocculation*; this is the process of breaking up aggregates into separate particles that are easily suspended and removed.

■ *Suspension:* Once insoluble particles are in solution, suspension will allow them to be flushed away, preventing them from settling and resulting in deposits.

■ *Peptizing:* This is similar to dispersion but more applicable to protein soils. This is the formation of solutions from soils that are only partially soluble.

■ *Rinsing*: This is the condition of solution or suspension that will allow them to be flushed from a surface. This is done by reducing water surface tension (wetting).

Water

Water used for sanitation must be potable whether it is from a private well or a municipal source. Potable means that it is fit for human consumption without further treatment [15]. Under no circumstances should nonpotable water be considered for cleaning. In addition, plant water lines should have backflow prevention or vacuum break devices installed to prevent backsiphonage of nonpotable into potable water systems. In all situations, sufficient boiler capacity is needed to provide enough hot water for the entire cleaning process. Cold water will not dissolve fats, so the plant must provide enough hot water to facilitate the entire sanitation process. Water hardness can have an impact on the effectiveness of cleaners and sanitizers as well as implications on the performance of plant equipment. All water contains some level of hardness due to minerals. Water hardness occurs when rainwater passes through the atmosphere and picks up levels of carbon dioxide (CO_2), creating a mild carbonic acid solution. Then as the moisture passes through the soil, it dissolves alkaline materials such as calcium and magnesium [4]. It is the minerals present in the water that results in hardness. Water hardness is typically measured in grains per gallon (gpg), and levels of hardness are described in Table 7.3 [4].

Some of the problems with hard water are reduction in effectiveness of cleaners and sanitizers, reduction in the effectiveness of heating equipment (i.e., boilers or

Table 7.3 Water Hardness Measurements

Hardness	Grains per Gallon (gpg)
Soft	0–3.5
Moderately hard	3.5–7.0
Hard	7.0–10.5
Very hard	>10.5

cookers) as scale forms on transfer surfaces, and contribution to the formation of biofilm on equipment. One means of softening the water is through the use of sequestering and chelating agents in cleaning systems to reduce hardness [3]. However, this can be more expensive than softening the water in the plant system. This can be done through the addition of chemicals (hydrated lime and soda ash) to precipitate the hardness. This is especially effective for boilers [4]. Another way to soften plant water is through ion exchange, in which sodium ions are exchanged for calcium and magnesium, making the water more compatible with cleaning solutions.

Cleaning Chemicals

When selecting cleaning chemicals and sanitizers, it is also important to match the type of material used for the processing equipment so that the equipment does not deteriorate. Soft metals such as aluminum can pit from harsh chemicals such as acids or unbuffered alkali, and the pitting can become harborage for bacteria and support the formation of biofilms. They must be safe for use and easily rinsed from equipment [3]. Common cleaners or detergents are listed in the following text:

■ Alkali: These are soil-displacing, emulsifying, saponifying, and peptizing agents. They also prevent mineral scale. These are some of the most commonly used detergent compounds and include sodium hydroxide and caustic soda (lye). They can be corrosive to aluminum and galvanized. Lye is especially strong and results in corrosion and is difficult to rinse. Trisodium phosphate (TSP) is milder, less corrosive, and does not precipitate hardness [2,4].
■ Phosphates: These emulsify and peptize soil and soften water and prevent soil deposits and mineral precipitation. They condition water and are a source of alkalinity [2]. The use of phosphates depends on hardness of the water.
■ Wetting/surfactants: These promote wetting or precipitate soil and rinse well [2]. They emulsify, disperse, and suspend soil such as oil as small droplets in water [4]. These agents are noncorrosive, soluble in cold water, and are not affected by harshness. They are stable in acid or alkali conditions. There are three types of such agents:
  – Anionic—pH neutral, compatible with acid or alkali.

- – Nonionic—better for oils, marginal effect of hard water, some will depress foam.
  - – Cationic—wetting agents, typically quaternary ammonia compounds; these do not react favorably with mineral soil but have some antimicrobial action.
- ■ Acids: These are good for cleaning alkali soils and removing minerals, especially calcium and magnesium, can condition water. There are two types:
  - – Inorganic—strong and corrosive, not recommended for food plant use. Examples are phosphoric acid, hydrochloric acid, and sulfuric and nitric acids.
  - – Organic—most useful in food plants as they are not as corrosive. Examples are acetic acid, tartaric acid, and lactic acid.
- ■ Chlorinated cleaners: These increase peptizing and minimize mineral stone as well as make alkaline cleaners more effective. This is not a sanitizing agent when used in a cleaner, so a separate sanitizing step is still required.
- ■ Chelating agents: Also known as *sequestering agents*, these control mineral deposits (such as calcium ions) through water softening, displace soil through peptizing. They are stable to heat and prevent precipitation of hardness. The most common is ethylene diamine tetra acetic acid (EDTA).

The sanitation chemical supplier should be a source of technical aid in the use and application of cleaning chemicals, not just a chemical salesman. The most effective supplier representative will conduct a plant survey to determine products, soils, equipment, facility flow, and personnel needs. They will participate in setting up the cleaning procedures but will not set up the procedures themselves. They will provide chemical safety training and training on application and cleaning technique, and should have sufficient microbiological knowledge to understand which organisms are of concern based on products made and to assist with selection of sanitizers to control these organisms. They should also provide an MSDS sheet for all cleaning and sanitizing compounds and verify that these materials are listed in the *Proprietary Substances and Non-Food Compound* publication. They will make sure that recommendations for any new compound is reviewed with and approved by QA and Sanitation. Their visits should be inclusive of chemical use and expenses, cleaning effectiveness, safety, and procedure review. All visits should be reflected in a report provided to the plant.

Cleaning Systems and Equipment

Central systems provide hot water and chemicals to stations placed around the plant. They can be automatically set to deliver various combinations of water pressure and volume for specific soils. The systems can be set to deliver chemicals directly into the water or be equipped with a mixing valve to add chemical from a central dispenser. The advantage to the mixing valve is that the same hose can be used for pre-rinse and final rinse of equipment with the valve off to prevent flow of chemical [15]. If this type of system is used, it is important to provide a sufficient number of hoses and hook up stations for the size of the plant and to have sufficient pump size to deliver the water or chemical needed for the distance and number of units to be run at any time. Spray guns for the cleaning system must be of sufficient size to deliver the chemical to the areas being cleaned. If equipped with a nozzle with a 15° spread, this should be sufficient for most applications [15]. Nozzles are available to alter the spray pattern for varying cleaning jobs.

An alternative to a central system is individual hose stations. These stations have a supply of hot and cold water or steam injection for creation of hot water. They require a gauge for direct reading of water temperature and will also be provided with individual backflow prevention devices [15]. Since they operate off city water pressure, the hoses will have adjustable nozzles for stream flow or mist spray. Hoses must be made of material that is able to handle the temperature and pressures for the system, be lightweight, flexible, and nonporous so that they can be cleaned.

Portable equipment can be used for wet cleaning of hard-to-reach areas. They may require hot water feed or have built-in heaters to create hot water. In addition, they can incorporate detergent and create foam.

Wet/dry vacuums are valuable in any food plant environment due to their versatility. They can be used for small cleaning jobs and are highly portable. Plants may also find value in purchasing floor scrubbers. Though they need space to operate, they are good for cleaning warehouses floors and aisles.

The Cleaning Process

The intent of the cleaning process is to remove solids and soils that have accumulated during the manufacturing process. The frequency with which the plant is cleaned will depend greatly on the operation and the types of soils involved. Dry mix plants are often continuous with ongoing cleanup or a full cleanup at the end of a week. Slaughter and further process plants now have flexibility from FSIS to determine what is appropriate provided there is no insanitary condition or no production of adulterated product. A rule of thumb for plants is to clean once within a 24-hour period and after the following product changeovers: between allergen ingredients for food safety, between animal species (i.e., when changing from chicken to beef to prevent economic adulteration), and between spices for quality reasons (i.e., from more colorful or flavorful spice blends to less intense blends). Consult with the plant sanitarians and QA department to determine the appropriate cleaning frequency. Decisions to clean equipment or rooms less than every 24 hours must be supported by sufficient scientific documentation to prove that this will not result in insanitary conditions and the production of adulterated product. This may come in the form of microbiological testing, scientific literature, or validation by a process authority. Environmental cleaning of areas such as wall, floor, ceiling, drain, etc., will be conducted daily or as often as necessary (i.e., during midshift cleanup, between shifts, etc.) to prevent product or contact-surface contamination. Nonproduction areas will be cleaned as frequently as needed to prevent transfer of insanitary conditions to production areas.

The following steps are basic procedures for effective cleaning and sanitizing. Each step in the process is reliant upon effective completion of the previous step.

The sanitation process starts with a dry pickup of scrap, paper, packaging, and product or ingredient spills. This should be ongoing during the manufacturing shift to prevent excess buildup and prevent insanitary processing conditions. Evening sanitarians can be assisted by having production personnel conduct the dry pickup at the end of their shift to save time and effort. The use of squeegees and shovels are preferred for floor cleaning over brooms or brushes. Squeegees are easier to clean than brooms, which can become encrusted with food material [12]. Avoid the use of air hoses to prevent blowing contaminants into product or product streams. Break down equipment to component parts or open equipment panels to clean inside. When breaking down equipment take care with the parts to ensure that they do not get damaged. Damage, such as scratches or pitting, can result in creation of bacterial niches. It is also important to ensure that the equipment parts are maintained in an order in which they may be reassembled or that they are kept with the specific machine. Some equipment operates better with the specific parts that came with it originally. Cover electrical panels or motors with plastic and secure the plastic to prevent forcing water into areas that may result in damage. Lock out/tag out moving equipment, such as blenders, to prevent injury (i.e., equipment someone may have to reach into). Do not place equipment parts on the ground (floor) or onto stairs or platforms where there is foot traffic, as this may only create more bacterial niches and result in recontamination of clean parts (Figure 7.2).

Instead, have racks or COP tanks for these parts as identified in the COP identified in Specialized Cleaning Procedures and Equipment.

Follow dry cleaning with a hot water rinse to break up fat, remove visible soils, and combine with mechanical action to prepare surfaces for cleaning. Water temperature for cleaning is very important, depending on the soils present. Generally, the water temperature used is approximately 5°F above the melt point temperature of fat [4]. This means that the recommended temperature will be between 130°F and 160°F. Since proteins denature and will bind to surfaces making removal more difficult, do not let the water temperature reach 185°F or above. The rinse process will proceed as follows:

- Rinse equipment or facilities from top to bottom so that soil moves from the equipment surface to the floor.
- Reduce water pressure to prevent atomization. It is recommend that the water be delivered at high-volume (9 gallons per minute or greater), low pressure (<600 psi) rather than high pressure (600–800 psi) water [15]. High water pressure can atomize contaminated soils into the air that will then settle on equipment. It can also drive contaminated moisture into sealed areas that can

Figure 7.2 Placement of parts on the floor during sanitation or at any time should be discouraged.

then come back out onto product surfaces or product. If there is a need to use higher pressure due to the nature of the soil, it is at this step that higher pressure should be used, before the application of detergent and in a controlled manner to reduce overspray as much as possible [13].

■ Avoid heavy spray on floors and into drains as this can create aerosols that can be a potential source of microorganism contamination from noncontact surfaces to contact surfaces.

Dry pickup and rinse are very important steps to remove as much soil as possible as most chemical cleaners are not designed to remove excess soils or gross contamination [12].

Wash all equipment and environmental surfaces with detergent and hot water. Here again, the water temperature is most effective between 130°F and 160°F, depending on the cleaning chemicals used. Cleaning may involve several steps, and cleaning compounds used may contain several ingredients, depending on the soils to be removed, hardness of the water, and prevention of formation of scale [5]. The chemicals used will be selected to control soils identified in Table 7.2. Chemical mixing, if not done by an automatic dispenser, is very important to cleaning effectiveness and cost control. Train sanitors to properly mix chemicals, and if verification of chemical strengths is required, use test strips, titration chemicals, or pH/conductivity probes available with some ATP devices. Ensure that they understand that excess chemical will not make cleaning easier; in fact, it may make it more difficult by harming equipment, leaving chemical film, and wasting money [7]. Conversely, too little chemical does not save money as it is not going to clean as effectively and will ultimately result in failures such as reduced shelf life and quality, microbiological problems, or regulatory Noncompliance Reports (NRs) or control action. A pumping system with pressurized air is an effective means of applying cleaning compounds, with creation of foam. Foaming cleaners are most effective when large areas need to be cleaned and they should be left on the equipment long enough to break down soils but not long enough to dry, making removal more difficult (Figure 7.3). Once the chemical has been applied to the surface, it will require contact time to penetrate and break up soil, but not so long that they begin to dry. Cleaning must be aggressive, and the use mechanical action is needed to remove soil and prevent buildup that can contribute to biofilm formation. As a rule, scrub contact surfaces on a daily basis and indirect surfaces such as frames at least once a week [13]. Unfortunately, some chemicals are represented as "no-scrub" and touted as being effective without mechanical action. Experience and scientific data has shown that scrubbing is required to prevent the formation of biofilms. Scrubbing, however, should not be so intense as to cause scratches or gouges in the surface being cleaned as these then become harborage niches for bacteria and begin the formation of biofilms. In this instance, softer scrub pads

Figure 7.3   A COP tank is convenient and saves sanitation time of having to clean each part of complex equipment.

Figure 7.4 A well-designed cleaning room facilitates sanitation and sanitary storage of equipment and utensils.

or soft-bristled brushes may be a better alternative to "green pads" (Figure 7.4). In no instance should steel wool or copper scrubbers be used as these are very coarse and can score equipment and facilitate the formation of biofilms that will require removal as described in Chapter 5.

During the process of rinsing and applying cleaning compounds, run conveyors or other equipment at slow speeds to ensure that all surfaces are contacted. The frequency of removal of belts for scrubbing and soaking must be identified by the plant in the master sanitation schedule. Cleaning must include the undersides of the belts, and if the conveyor guides have removable UHMW runners, identify the frequency with which they need to be removed for cleaning and soaking. Do not allow sanitation employees to stand on top of conveyors or product surfaces as their boots may be a source of contamination that can be ground into equipment. Provide them with the appropriate ladders or lifts to reach high spots.

Once the cleaning process is completed, the rinsing process will remove soils suspended in detergent. Rinse all surfaces with hot (130°F–160°F) water to remove all soap. It is beneficial to use water at the lower end of this temperature range to prevent the creation of steam and the formation of condensation on overhead structures []. Use care to avoid overspray or water spray into floor and drain that might result in the creation of aerosols.

Specialized Cleaning Procedures and Equipment

COP is conducted when there are many small parts that come off a piece of equipment. They may be placed on a rack for cleaning or into a COP tank. The advantage of the COP tank is that the parts may be cleaned without a sanitarian handling each one and that can translate into time and labor savings. Ideally, the vat or tank is heavy-gauge stainless steel [15] and sufficient size to fully submerge all parts. It will also not be a source of contamination itself and will have smooth welds and no dead spots.

After dry-cleaning major soil off the parts, place them in the tank. Water added to the tank should either be hot (125°F–130°F) or the tank should have steam injected to achieve that temperature. Create turbulence, either from steam or mechanical means, to aid in loosening soil. Add the chemical cleaner determined to be most effective for the soil and parts and start the cleaning cycle. When parts are clean, rinse them thoroughly with clear potable water, inspect each part, and apply sanitizer. Cleaned and sanitized parts may either be reassembled or stored on a rack until ready for use.

A good place to locate the COP tank is in a designated cleaning room. This room will be made of materials that can handle a wet environment and the use of cleaning chemicals. It will have adequate drainage to prevent pooling of water and well lit so that the sanitarian can see the equipment being cleaned. The room will provide all the necessary tools for cleaning, such as hoses and hangers, chemical tanks or access to central cleaning system, storage for brushes or scrub pads, and racks or shelves to hold cleaned parts or utensils. Racks or hangers will be of a material that will facilitate drying of the cleaning utensils.

The room will also have adequate ventilation to pull off steam generated from cleaning. Another example of COP automation that may be beneficial is a pass-through basket or tote washer. Dry-cleaned totes or baskets are conveyed through the system where they are rinsed, cleaned, and rinsed again. They can then be stacked for inspection and sanitizing. Cost, space, and effectiveness are factors to consider when deciding whether to use this type of system.

CIP is often found in processes where liquid or flow-type materials such as juice and dairy products are being manufactured or where brines are prepared for meat injection or tumbling. This type of automated cleaning is usually a closed process system involving large tanks, kettles, or piping systems where there are smooth surfaces. It follows the basic cleaning process involving pre-rinse, soap, rinse, and sanitize [15]. CIP involves circulation of detergent through equipment by use of a spray ball or spray nozzle to create turbulence and remove soil. It is effective at removing soils and cost-efficient because it requires less labor and, where an automated dispensing system is used, very effective at containing chemical costs, minimizing labor, and reducing water and energy costs. The typical CIP system follows the basic steps for cleaning and sanitizing. There is an initial phase with removal of any small parts that cannot be cleaned within the system followed by a cool temperature (<80°F) rinse phase. Rinse water is flushed from the system, and the cleaning solution is circulated throughout for the time required to remove soil. There is a final rinse followed by the application of sanitizer [19].

There are several requirements for CIP systems to ensure that they are effective:

- The chemical solution must be capable of reaching all surfaces, and the surfaces are ideally made of stainless steel, not softer metals.
- The internal surfaces are round or tubular, not flat, and there are no ledges or recesses to prevent accumulation of soil that cannot be removed.
- Sanitary design includes smooth and continuous welds.
- The vessel is self-draining to remove all cleaning and sanitizing chemicals.
- Pump sizes are sufficient for the size of the tank or length of pipes to be cleaned. The rule of thumb is that the pump can produce a flow rate 4–5 times the rate of the product flow. The flow rate is recommended at 5 feet per second to achieve a scrubbing effect. To calculate the flow rate needed for sufficient volumes of cleaner, multiply the tank circumference times two. This will provide the minimum flow required in gallons per minute [19].
- The system is run by computer, in a prescribed manner, to control the flow, mixing and diversion, temperature, and time of the chemicals for cleaning and sanitizing.
- There are a sufficient number of tanks for the various solutions used, and they can contain sufficient quantity, about 50% more solution, than required.
- It is recommended that cleaning solution be changed approximately every 48 hours. Foaming-type solutions are not recommended for this application.

The plant should have a written program to monitor temperature, flow rates, or velocity in open systems, pressure in closed systems, and cycle times. This program shall include procedures and frequencies for monitoring. In addition, it is wise to physically evaluate the conditions of components, for example, the spray balls, to ensure that they are not damaged or plugged from mineral, thus affecting their ability to generate sufficient spray [19]. Monitoring records for the CIP process and inspection must be maintained to document findings and determine whether corrective or continuous improvement actions are required. It is extremely important that the CIP process not be shortcut or manipulated once the timing and flow rates are established. Changes in chemical concentration, timing, or flows can result in insufficient cleaning which in turn can allow the creation of biofilms. This will then require efforts to reduce the presence of films through breakdown, scrubbing, or the use of peracetic acid, which has been shown to be effective against biofilm (Figure 7.5) [23].

Equipment Baking

This has become a specialized practice in some RTE meat and poultry plants as a supplement to cleaning and sanitizing. In this process, product contact equipment

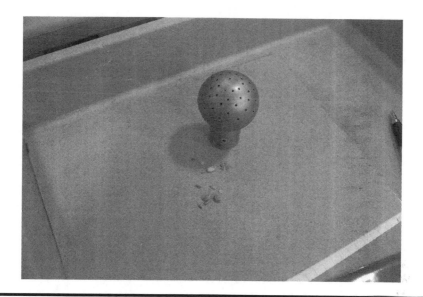

Figure 7.5   CIP spray balls will require periodic inspection to determine whether they are plugged or damaged, thus limiting effectiveness.

such as sausage peelers, slicers, racks, and tables are cleaned and then broken down to remove any parts that would be negatively impacted by high temperature (i.e., motors, gaskets, soft metals, etc.). The equipment is then placed in a smokehouse and cooked to raise the surface temperature to a lethal level of approximately 160°F for approximately 20–30 minutes [16]. It also heats internal and hard-to-reach areas for microbiological kill. This has proved to be an effective means of control for both pathogens and spoilage organisms.

Deep Cleaning

This is a specialized cleaning concept that is similar to baking. Again, complex equipment is taken apart to clean all components and remove as much soil as possible. Once this is done, dry steam is used to heat the metal to 180°F–185°F and penetrate deep into cracks, crevices, and pores. It is effective because steam under pressure is hotter and drier and penetrates better than water and larger-molecule chemicals [10]. It is important to remember that the equipment must be cleaned first; otherwise, the steam will simply bake on soils. Understand that this is only effective if the surface is brought to the 180°F–185°F temperature to kill microorganisms.

Floors and Floor Drain Cleaning

Floors and drains present a special challenge to sanitation as well as to manufacturing, especially in cooked RTE meat and poultry plants. Plant floors are exposed to most sources of contamination in the food plant starting with shoes, unless the plant provides captive shoes and/or uses floor foamers. They are exposed to raw materials or food products, heat and cold, and harsh cleaning chemicals. They are also subject to the movement of materials in vats or on pallet jacks or forklifts, and subject to moving or vibrating equipment. Because they take a lot of abuse, they are hard to keep sealed and nonporous, and that makes them good locations for microbial harborage [6]. They require special cleaning procedures to eliminate soil, bacteria, and biofilms that can harbor in cracks and crevices. Dry cleaning is the first step to remove heavy soils from ingredient or food products. To accomplish this, squeegees and shovels are recommended as bristled brooms are not as effective at moving wet or fatty materials. In dry areas, however, brooms and shovels may be appropriate. Once the majority of the heavy soils are removed, rinse the floor with low-pressure and high-volume water. It is important that soils and bacteria on the floor are not sprayed onto equipment or atomized through the use of high-pressure water spray. Follow this with the application of a cleaner that will break up soils on the floor. It is also effective to scrub the floor with a mechanical scrubber or stiff-bristled brush to help break up biofilms. The final rinse to remove all detergent and suspended soil will be at high volume but low pressure to prevent aerosols that can carry bacteria. Squeegee or vacuum the floor to remove as much standing water as possible, and then sanitize the floor at a higher level than for product contact surfaces, such as 800–1000 ppm.

Floor drain cleaning procedures are especially important in RTE meat and poultry plants where *Listeria monocytogenes* is a concern, since the organism grows very well in floor drains. This is not a problem if the organism stays in the drain; however, occasional backups occur, and at times, spray from hoses is directed into the drain, atomizing bacteria-laden moisture. Begin cleaning drains by opening the covers and removing the strainer baskets and quat blocks or rings. Remove any debris from the drain that may have gotten through the strainer, and then pre-rinse with clear warm water at low pressure. Apply foam cleaner to drain and around the drain. Powdered caustic cleaner is very effective both in and around the drain. Scrub with a brush specifically designated for floor drains, as deep into the drain as possible. Following the scrubbing process, rinse with low-pressure potable water, and inspect prior to sanitizing. As with the floor, sanitizing with a higher level, 800–1000 ppm is recommended. Drains usually have P-traps to prevent backup of sewer gases, so it is also recommended that the water in the trap be replaced with an equal amount of sanitizer. This usually requires about a gallon of sanitizer poured into each drain trap.

Brushes used for floors and drains must be identified as such and not used for any other cleaning process to prevent cross-contamination from these areas. As a practice, they should also be kept separate from brushes or items used to scrub equipment surfaces. As an added measure of cross-contamination prevention, separate brushes used for raw and RTE product area drains and separate brusher used for raw and RTE equipment surfaces.

All cleaning equipment and utensils must be maintained in a sanitary condition. Squeegees, brushes, and floor scrubbers should be cleaned to prevent accumulation of soils that will reduce their cleaning effectiveness. They should be stored in a manner that will prevent their contamination and allow them to dry between uses.

Sanitizing

Where cleaning is intended to remove solids and soils such as fats or protein, sanitizing is intended to reduce microorganisms, specifically those of public health concern or those that can cause product spoilage. Sanitizing is different from disinfection. Sanitizing will reduce the bacteria to levels that are considered, whereas disinfection will destroy or irreversibly inactivate infectious fungi and bacteria. It is more common to sanitize in the food industry as opposed to disinfection. [21]

There are two steps that need to take place prior to the application of sanitizer to clean surfaces. The first is to remove all condensation from overhead surfaces. This is best done by blotting rather than wiping, which can cause the condensate to drop onto clean surfaces. Several methods are identified in Chapter 6 to control and prevent condensation. There are several effective tools that can be used to remove condensate when it forms. For ceilings, extension poles with sponges or paper towels or even unused paint rollers that have been dipped in sanitizer. As with all cleaning utensils, these must not come in contact with the floor and should be stored to prevent their contamination, either on wall-mounted holders or in sanitizer buckets, where they will be accessible for use to blot condensation if it forms during operations. The second step is to conduct pre-op inspection of food-contact surfaces to ensure that they are free of visible food particulate and soils. Inspection is a critical element to sanitation as it is your assurance that your procedures are effective and have been properly administered. It also required as part of the SSOP plan. It begins before sanitation is even completed by observing sanitarian practices and providing them with guidance on technique. Once they have completed their procedures, have sanitarians inspect equipment before QA conducts pre-op or with QA as part of pre-op so that they can be shown the tough spots to clean and know where contamination can accumulate.

Once cleaning is complete, and prior to application of sanitizer, it is a good idea to conduct basic monitoring, beginning with organoleptic inspection:

- Look in, around, and under equipment and structures for indications of soil removal.
- Smell the environment. Does it smell clean or are there sour or musty odors?
- Feel equipment surfaces for grease or grit from incomplete soil removal.

Provide the proper tools to monitor, including a flashlight; mirror (no glass) for inspecting difficult to reach locations; test strips or kits to monitor cleaning and sanitizing solutions; thermometer for checking water temperature; a ladder or lift to inspect high equipment and overhead structures; and a notepad and pen to record findings. Inspection findings should be noted as they are observed, clearly describing the condition, detailed corrective actions, and effectiveness of the corrective actions. These findings should be tracked to determine whether trends are occurring. It is far better for the plant to identify trends and take corrective/preventive actions, that for FSIS to identify poor sanitation trends.

Although this will be covered in greater detail in Chapter 8, it is very important that sanitarians be trained to conduct inspection of their work and this inspection should also incorporate the use of ATP Bioluminescence technology to verify cleaning effectiveness. This technology does not indicate the presence of microorganisms; rather, it measures the presence of organic material, indicating an environment in which microbes can live and grow. Thus, if used in conjunction with visual inspection, it provides sanitarians with immediate feedback as to the efficacy of their cleaning efforts.

The final step in the process is to apply sanitizer to all cleaned and thoroughly rinsed surfaces to destroy hidden microorganisms. Effective use of sanitizers is integral to controlling microorganisms for the purpose of food safety and product shelf stability. It must be understood that sanitizing does not replace thorough hand washing or equipment and facility cleaning. Factors such as presence of organic materials, especially protein, decrease the effectiveness of sanitizers against microorganisms. Therefore, plant employees must be trained to properly wash hands and clean equipment/facilities, to remove all carbohydrate, fat, and protein soil and biofilms prior to sanitizing. The sanitizing step will supplement effective cleaning through the reduction of microorganisms to a level considered safe. This differs from disinfection, which is the complete removal of pathogens and reduction to the lowest level of other microorganisms.

The selection of sanitizers will be determined to a great extent by the microorganisms to be controlled in the operation, so it is very important to know what organisms are of concern in your specific operation. Other factors to consider when selecting a sanitizer are corrosiveness to the equipment present in the plant and the cost of application at the concentration needed. Table 7.4 identifies many of the common sanitizers for food plant use. An effective sanitizer will pass an efficacy test, which requires that 99.999% of harmful microorganisms (roughly a 5-log reduction) be killed within 30 s [1]. All sanitizers used in the plant must be listed in the USDA Approved Chemical Compound book, must be US EPA approved, and an MSDS must be provided for each sanitizer and accessible to employees. Sanitation employees handling sanitizers must be trained to properly handle and prepare sanitizers at the appropriate effectiveness level and at the level specified by the manufacturer. They must provide with strips or chemical titration kits to test the concentration level of the sanitizer. They must also be trained to clearly identify any container bearing sanitizer with the type of sanitizer it contains for safety and regulatory reasons.

If the sanitation effort is effective, sanitizing will give an extra measure of microbiological control. The environment to be cleaned and the organisms of concern

Table 7.4 Food Plant Sanitizers

Sanitizer	Organisms Controlled	Usage Level without Rinse	Residual	Corrosiveness	Stability
Chlorine (sodium hypochlorite)	Gram −, spores, bacteriophage. Quick kill.	200 ppm without rinse.	None	Very corrosive especially to soft metals. Pungent.	Unstable in hot water, presence of organics.
Quat (quaternary ammonium)	Gram +, inhibits mold. Not effective on spores, fungi, or bacteriophage.	400 ppm without rinse.	Slight	Minimal.	Stable even at high temperatures and wide pH. Affected by some mineral content.
Iodophores (Iodine and stabilizing agent)	Gram + and Gram − but not effective against spores or bacteriophage.	25 ppm with no rinse required on contact surfaces.	Slight	Minimal (except for galvanized) and nonirritating. Can stain belts and PVC.	Stable below 120°F but loses stability between 120°F and 140°F.
Ozone	Gram +, Gram −, viruses, protozoa. Effective against biofilms.	Can be applied directly to foods.	None	Mild.	Unstable, quickly breaks down.
Peracetic acid, peroxyacetic acid	Gram+, Gram−, wide spectrum but not effective against spores. Quick kill.		None	Corrosive to soft metals, but not stainless. Pungent.	Stable, but impacted by organic load and neutral pH.
Chlorine dioxide [8]	Wide spectrum. Very quick kill. Can penetrate biofilm.		None		Breaks down to water, oxygen, and NaCl.

will again dictate selection of sanitizers. It is also recommended that the maximum amount of sanitizer at a no-rinse level be applied to surfaces for the maximum effect.

Some of the more commonly used, and effective, sanitizers are discussed in greater detail below.

As indicated in Table 7.4, sodium hypochlorite (chlorine) is most effective against Gram-negative bacteria (*Salmonella*) and one of the least expensive sanitizers. It kills through damage to the outer cell membrane, inhibition of cellular enzymes, and destruction of the DNA in microorganisms, but spores are resistant. It is unstable in warmer water and can be hard on some metals if used exclusively. It is also impacted by impurities such as mineral and organic material in water. Use of this sanitizer requires the application of break-point chlorination. As chlorine is added to water, it is bound by the impurities in the water up to a point. Once chlorine demand of the impurities is met, the break point, the amount added beyond this point is free residual and available for sanitizing surfaces [4].

Quaternary ammonia (quat) kills a wide range of organisms and is most effective against Gram-positive bacteria (*Listeria*). It kills by blocking the uptake of nutrients and preventing the discharge of waste [21]. Quat is much more stable in warm water, has minimal odor, and is not staining. It is the preferred primary sanitizer in cooked RTE meat and poultry facilities for equipment and environmental surfaces as it is most effective against *L. monocytogenes*. It is not as effective against Salmonella [17]. The maximum level of quat permitted on product contact surface, without rinse, is 400 ppm; however, because it has a residual, it is also a good idea to apply quat after cleaning on the last day of the week at a level of 800–1000 ppm and without rinsing off [18]. Prior to the next start of operations, rinse product contact surfaces, and reapply quat at 400 ppm with no further rinse. Sanitize walls, floors, drains, and overhead structures (e.g., air units) with 800–1000 ppm quat, and use quat in shoe sanitizer mats or floor foamers at a level of 800–1000 ppm.

Chlorine dioxide is an inorganic sanitizer that has a broad range of bacterial kill, including viruses and fungi. It is an oxidizer that reacts with proteins and fatty acids within the cell membrane, causing loss of permeability control and disruption of protein synthesis. Some of the advantages of chlorine dioxide are that it is safe in solution and is effective at levels as low as 5 ppm when applied for a minute. It can be used for disinfection at levels of 100 ppm when left on for approximately 10 minutes. It is effective in the presence of organic material, but as the concentration the efficacy is decreased. It is considered environmentally friendly, and it does not form chlorinated organic compounds [21].

Peracetic acid (PAA) kills a broad range of microorganisms and spores by disrupting the chemical bonds in the cell membrane. PAA can be used effectively in cold conditions, which makes it desirable for the control of *L. monocytogenes*. It is safe for use when safety precautions are taken (use of personal protective equipment) and is environmentally friendly. However, it can be impacted by a high organic load and loses efficacy as the pH approaches neutral [21].

It is recommended that plants alternate sanitizers during the week to prevent proliferation of specific flora [5]. Since some sanitizers are more effective against Gram-positive organisms, using them exclusively can eliminate the Gram-positive organisms but will eliminate the competition for the Gram-negative organisms, and they might flourish. So alternating sanitizers will prevent the elimination of one organism only to allow the proliferation of another. As an example, in a five-day production week, use quat on four days and chlorine on one day. The progression of alternating sanitizers would be: quat on Monday and Tuesday, chlorine Wednesday, quat Thursday and Friday, and on the last day of the week if weekend work is conducted. The use of chlorine one day per week will not deteriorate equipment if it is applied at appropriate levels. Alternating quat and chlorine to maximize bacteria-killing effect is a good idea; however, for safety reasons, never mix quat and chlorine together as they can produce a dangerous reaction and a toxic gas.

There has also been discussion about sanitizer rotation as a means of preventing bacterial adaptation and eventual resistance. The process of genetic mutation takes many generations, and this is not the point of discussion in this instance. However, if the effective kill level of the sanitizing process is 99.999%, then the higher the initial load, the greater the number of bacteria that will be present after sanitizing. Some of the survivors may be naturally resistant to the sanitizer, or they may be protected by biofilm. As these survivors multiply, there is a possibility that there will be a proliferation of bacteria that are resistant. In this instance, it may be necessary to change sanitizers to one that will effectively kill the survivors or to disinfect [21].

Fogging with quat or chlorine can be an effective means of getting sanitizer into pores and crevices. However, do not fog facilities with quat or chlorine while personnel are present, as the mist is highly irritating. People must be thoroughly trained to conduct fogging, and areas must be secured to prevent access while fogging, and protective gear must be provided.

Ozone

Ozone is gaining wide acceptance in the food industry as a primary sanitizer or as an alternate sanitizer where a "multiple hurdle" sanitizing approach is used. We often associate ozone with the clean air smell after a thunderstorm, when electrical charges pass through the air and create ozone gas. It can also be created in food manufacturing plants by passing high-voltage electricity through air creating a triatomic form of oxygen (O_3). It has been used in Europe for purifying drinking water and is used in the United States to purify city water in many areas. In June 2001, the U.S. Food and Drug Administration officially granted GRAS (Generally Recognized As Safe) status to ozone for use in food contact applications, and in December 2001, the USDA approved the use of ozone for contact with meat and poultry products [11].

Ozone is an extremely strong oxidizer, and this is how it disinfects. This is one of the most important factors in its effectiveness. It works against a very wide range of organisms, both Gram-positive and Gram-negative viruses and protozoa, and because it works as an oxidizer of bacterial cells, they do not develop resistance as they may with other sanitizers. There are several other attractive advantages to ozone versus other sanitizers. First, it is generated as needed using specially designed equipment; therefore, the plant does not need to store sanitizing chemicals on premise. Ozone is not a very stable molecule and ultimately breaks down and releases the additional oxygen atom, and this is important because it means that it can be applied directly to contact surfaces without the need for a rinse. This is also important because, although it does not leave a residual like quat, it is more environmentally friendly. Second, as it breaks down, it does not add to the biological oxygen demand (BOD) or the chemical oxygen demand (COD) in plant effluent, which can reduce treatment surcharge if the plant is on a city system. Finally, because it is approved for direct product contact, it is also effective as rinse water for vegetables and can be misted onto slicer blades or other equipment handling RTE products.

One caveat is that because ozone is a powerful oxidizer, it is important to monitor the amount of gas that may be present in the environment and to protect employees from potential lung irritation. While it can kill organisms in the air, and has been used successfully in produce operations to extend shelf life, it should not be used as an air treatment unless there are no people present at the time and there is time to air out the plant before people enter.

Operational Sanitation

It is very important to maintain a plant operational environment that will ensure the production of safe food products during operations and the prevention of conditions that may lead to product contamination. This is part of the requirement of the SSOPs as described in Chapter 1. While this is a regulatory requirement, it should also be an objective of the plant to ensure that product does not become contaminated from the effects of operational processes.

- GMPs: These are covered in great detail in Chapter 9 and are possibly the most important factor in the maintenance of sanitary operating conditions. This includes employee dress, hand washing and maintenance of hand-wash facilities (providing hot water, soap, and towels), prevention of raw to cooked employee traffic, food and tobacco control, and disease control. These controls must be enforced with plant visitors as well.
- Spill control: All areas within the building, storage and office areas, as well as processing areas, must be kept clean, neat, and free of soil or spills. Ingredient spills in storage or manufacturing must be cleaned up immediately. Excess accumulation of spilled product, ingredient, or other material (i.e., grease) in

or around where product is produced must be cleaned up for food safety and employee safety.

- Operational soil buildup: Remove buildup of food soils on equipment, especially in product flow zones to prevent possible bacterial growth. Single-use, disposable towels are preferred to multiple use towels for wiping down equipment surfaces. Towels used should be wetted with sanitizer before use to provide an additional microbiological control benefit.

- Trash/inedible control: Inside trash or recycling bins will be kept closed with lids secured or emptied frequently to prevent overflow. Inedible carts will be emptied frequently and cleaned no less than daily. No trash or solid waste will be stored within the plant except in covered containers. Waste containers at packaging machines may be uncovered during the operation of such equipment.

- Idle equipment control: Idle equipment, not currently in use, will be covered to prevent contamination from ongoing operations, maintenance, or sanitation activities. If not covered, equipment will be washed on a daily basis and, in all cases, washed and sanitized prior to use in production.

- Floor mats or foamers: Foamers are the preferred means of delivering sanitizer on the floors. They will be maintained to deliver quat at 800–1000 ppm and set to create thick foam rather than a liquid pool. If floor mats or baths are used, they must be maintained at optimum sanitizer strength. Use 800–1000 ppm quat for all floor mats.

- Overhead control: Overhead structures (i.e., pipes, beams, lights, etc.) are to be clean, free of condensation as well as dust buildup, rust, and flaking materials (paint, silicone, tape, plastic). Place overheads on the master sanitation schedule to prevent buildup of soils that can drop onto product surfaces or product.

- Packaging control: Direct product contact packaging (i.e., film, bags) will be covered while in storage areas to prevent accidental contamination. Packaging materials returned to storage will be dry, clean, and sealed.

- Door control: Doors (including silos, compactor areas) will remain closed during operations unless they have operating air curtains to prevent the entry of pests and outdoor dust or soil.

- Maintenance control: Food contact equipment contaminated by maintenance activity before or during operations will be properly cleaned and sanitized before contact with food product or packaging.

Mid-Shift Cleanup

Many meat and poultry product plants conduct mid-shift or between-shift cleaning processes. This involves a stoppage of work and follows the basic sanitation process and use of sanitarians. It may be worthwhile for a plant to evaluate the need for a thorough sanitation process at mid-shift by conducting a microbiological monitoring of indicator organisms over the course of production. If there is an indication that a dry cleaning is sufficient, there may be a savings of labor, chemical, and energy (hot water) dollars. In addition, it is always desirable from a microbiological growth standpoint to keep the process as dry as possible. If it is determined that wet cleaning is necessary, do not clean while other lines are running unless they are screened or otherwise segregated. Do all that you can to avoid overspray from lines being rinsed to clean lines, the creation of condensate from overhead surfaces and especially aerosols from spraying water into drains.

Inspections

As stated earlier in this chapter, the plant manager is responsible for maintenance of sanitary conditions at his plant. As such, the plant manager shall make regular routine operational sanitation tours of his location to determine the effectiveness and adequacy of local housekeeping programs. Each plant is required to make a documented daily operational sanitation inspection to fulfill requirements of SSOPs. However, at least once per quarter, the plant manager should conduct a review of the facilities and grounds, evaluating operational sanitation, maintenance, food protection, and Good Manufacturing Practice (GMP) conformance. This review should be made with a cross-functional plant team, composed at a minimum of the following personnel; plant manager, QA manager, maintenance manager, and at least one hourly employee (preferably a lead person). Those items that can be corrected at the time of the review will be corrected and correction documented on a review report. Those items that cannot be corrected at that time should be noted on the report for follow-up correction with an assignment to an individual or department and a timeline for completion. The team should also evaluate the FSIS noncompliance report (NR) results to correlate with their findings, in the event there are any repeated deficiencies, and to identify opportunities for improvement to reduce the frequency of NRs.

An effective sanitation program takes in many factors: sanitarian selection and training, effective procedures for daily and regularly scheduled cleaning, selection of the right chemicals for cleaning and sanitizing, implementation of standard cleaning processes, and maintenance of operational sanitation conditions. In addition, plants should evaluate inspection reports, microbiological results and, as applicable, NRs to periodically assess sanitation performance. The objective is to drive continuous improvement in the sanitation process, whether it be retraining of sanitors, rewriting the sanitation procedures, or redesigning plant equipment for more effective cleaning. If all of these processes are implemented, they will go a long way in ensuring the plant of the safety, wholesomeness, and quality of their products.

References

1. Stauffer, John E., *Quality Assurance of Food, Ingredients, Processing and Distribution*, Food and Nutrition Press, Trumbull, 1994, p. 132.
2. Gould, Wilbur A., *CGMP's/Food Plant Sanitation*, CTI Publications, Baltimore, 1994.
3. Troller, John A., *Sanitation for Food Processing*, Academic Press, 1983, chap. 6.
4. Katsuyama, Allen M., *Principles of Food Processing Sanitation*, Food Processors Institute, 1993.
5. Gregerson, John, Clean Sweep Sanitizers and Disinfectants Are Only as Good as the SSOP's That Govern Them, *Meat Marketing and Technology*, January 2005, pp. 53–58.
6. Castaldo, Dr. D. J., The Enemy Below, Cleaning and Sanitizing Floors Present Unique Challenges for Controlling Pathogens and Other Filth, *Meat Processing*, February 2005, pp. 32–36.
7. Anon, When It Comes to Sanitation, Training Makes the Difference, *Food Safety Magazine*, 2005.
8. Anon, 10 Reasons Why You Should Be Using Chlorine Dioxide, *Food Safety Magazine*, 2005.
9. Carlsberg, Henry, Can Sanitation Be Considered Technical?, *Food Quality*, March 2005, p. 76.
10. Bjerklie, Stave, Going Deep, US Meat Processors are Beginning to Explore Deep Cleaning Equipment with Dry Heat, *Meat Processing*, April 2004, pp. 34–36.
11. Rice, Dr. Rip G., Dee M. Graham, and Matt T. Lowe, Recent Ozone Applications in Food Processing and Sanitation, *Food Safety Magazine*, October/November 2002, pp. 10–17.
12. Stier, Richard F., Cleanliness Is Next to Godliness and Essential to Food Safety, *Food Safety Magazine*, April/May 2004, pp. 30–65.
13. Redman, Rory, Basic Elements of Effective Food Plant Sanitation, *Food Safety Magazine*, April/May, 2005.
14. Frank, Hanns K., *Dictionary of Food Microbiology*, Technomic Publishing Company, Lancaster, 1992, p. 94.
15. Imholte, Thomas J. and Tammy Imholte-Tauscher, *Engineering for Food Safety and Sanitation*, Technical Institute of Food Safety, Medfield, 1999, pp. 267–270.
16. Tompkin, R. B., Control of *Listeria monocytogenes* in the Food Processing Environment, *Journal of Food Protection*, Vol. 65, No. 4, 2002, p. 720.
17. Mustapha, A. and M. B. Liewen, Destruction of *Listeria monocytogenes* by Sodium Hypochlorite and Quaternary Ammonium Sanitizers, *Journal of Food Protection*, Vol. 52, No. 5, 1989, p. 306.
18. Anon, *Guidelines to Prevent Post-Processing Contamination from* Listeria monocytogenes, National Food Processors Association, April 1999, p. 9.
19. Stier, Richard and Michael Cramer, Top Tips to Make Your CIP and COP Systems Work for You, *Food Safety Magazine*, The Target Group, Glendale, 2005, pp. 14–18.
20. Head, OSHA Compliance and Educational Product Development, AIB International, *QA Magazine*, July–August 2012, pp. 6–8.
21. Funtner, Alan P., Sanitizers and Disinfectants: The Chemicals of Prevention, *Food Safety Magazine*, The Target Group, Glendale, 2011, pp. 16–19, 77.
22. Fuhrman, Elizabeth, Sanitized Properly, *The National Provisioner*, 2012, pp. 74–77.
23. Pellegrini, Meg, Automizing Safety, *The National Provisioner*, 2012, pp. 72–74.

Chapter 8

Verification of Sanitation

Now that we have got low micro results, can we cut down on the testing to save money?

Anonymous (former) plant manager

Once the sanitation process has been completed, everything in the plant is ready for production to begin, and there is no further action to be taken, correct? As expressed in a car commercial from a few years ago, "Not Exactly!" Equipment may appear to be clean and free of gross soils and contamination; however, there may be organic buildup and microorganisms not visible on the surface or hidden in niches. Assurance that sanitation is effective requires a monitoring system that encompasses steps involving verification and validation. The processes of verifying and validating sanitation are very different. In verifying, you are making an immediate determination that sanitation was effective for that prior process. However, with validation you are determining that the process is effective over a period of time.

Verification

No sanitation system would be complete without means of verifying effectiveness. Verification may be done in several ways, from simple and relatively inexpensive to slightly more expensive and complex. The least expensive and easiest to implement is visual or organoleptic examination of the postsanitation and preoperation environment. Organoleptic pre-op inspection is required as part of the plant SSOP, and there is no regulatory requirement to incorporate other investigative tools.

However, added investigative tools and documentation provide extra insight into the thoroughness of the sanitation process. ATP or bioluminescence measurement is an extremely effective, relatively inexpensive tool that many food manufacturers employ for rapid verification feedback about sanitation. Microbiological testing is a tool used by many companies as a means of both verification and validation of sanitation. Several verification methods will be explored in this section with the intent of addressing the decision-making process on the type and volume of tests conducted. Documentation of sanitation verification findings will be presented as a means of tying together the elements of sanitation with the other critical components of an integrated food safety system.

Organoleptic

Organoleptic verification involves all of the senses with the exceptions of taste and hearing. This verification is typically done during pre-op inspection following the sanitation process. It involves inspection of the equipment obviously using the sense of sight to look for any indications of food material left on equipment, such as grease, dough, or produce, depending on the products being handled in the plant (see Figure 8.1).

To facilitate inspection, the processing areas should have sufficient lighting, as identified in Chapter 6. However, this lighting may be supplemented with the use

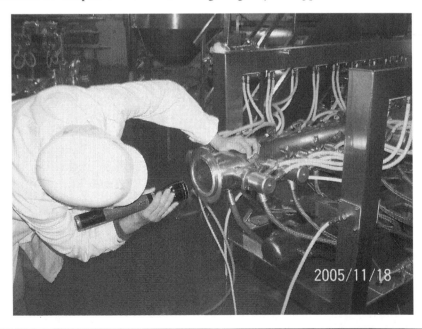

Figure 8.1 Verification of sanitation starts with an initial visual evaluation of equipment to verify cleanliness.

of a flashlight for areas that are semienclosed or where lighting is not sufficient. Other tools that may be of value to pre-op inspectors are a mirror (polished stainless, not glass) on an extendable rod to look at the undersides of equipment and avoid excessive bending. A ladder or lift may be utilized to look at high areas such as overhead pipes and refrigeration units. When using a ladder or a lift, always follow recommended safety practices to avoid injury. During pre-op inspection, other senses will also be employed. Touch the equipment to find any greasy residue that may be from food or grit that may be the result of inadequate soil removal. Look and feel for the development of slimy surfaces that may indicate the growth of bacteria [20]. Smell the plant and around equipment for any sour or musty odors that may be coming from niches that are not easily or frequently cleaned and that may be harboring spoilage bacterial growth.

The plant must take primary responsibility for pre-op inspection that identifies needs for corrective action following sanitation. There is a problem when the plant relies on the USDA inspectors to find sanitation problems and point them out to the plant. Training of sanitation inspectors allows the plant to discover deficiencies and take corrective actions and avoids the "Bucket Brigade" of sanitation personnel following the inspector and correcting deficiencies he or she finds. An effective inspection process relies on the sanitation department to monitor their own work with QA providing verification and release of a department before USDA conducts their inspection. It is a good idea to develop a system to notify the USDA inspection when the plant is released for them to conduct their regulatory inspection. Some plants employ a tagging system to notify USDA and production when an area has been inspected and released by sanitation and QA. A "Released by QA" tag is hung, providing an indication that the area is ready for USDA inspection or production setup to begin. Production and/or maintenance personnel must be trained not to enter a production area to begin setup of equipment until sanitation and QA have had sufficient time to conduct their inspection. This training must be enforced for pre-op inspection to be effective. It is generally understood by QA personnel that production departments need to start on time to be efficient and meet production goals; however, cleaning and inspection time must be built into the manufacturing time frame to ensure that QA has time for inspection to verify the plant has a clean and safe environment for production.

Documentation of pre-op inspection should be prepared real time; that is, while the inspectors are on the floor and immediately during the observation of a finding. It is not a good idea to take the report back to an office or break room to record the results. Reports will be completed while findings are fresh in the inspector's mind, so they can provide accurate detail of the finding and the action taken by the plant. They will include information pertaining to the sanitation deficiency, product control action, actions to restore sanitary condition (i.e., recleaning and sanitizing), and actions taken to prevent reoccurrence. As identified in Chapter 1, pre-op sanitation records will be maintained on site for at least 48 hours before they are moved to an off-site location. However, they must be made available to inspection personnel in a timely fashion when they are requested. They must be retained for a minimum of six months, but it is recommended that they be retained for a year past the shelf life of the product. It is acceptable to maintain records on computer provided proper security will prevent altering or tampering.

ATP Bioluminescence

Bioluminescence is a technology that has been used successfully by food companies for several years now. The science behind bioluminescence is based on the chemical adenosine triphosphate, or ATP. All living cells contain ATP as part of their cellular makeup. ATP powers energy-consuming reactions [20]. Microorganisms contain ATP, as do food products from nonmicrobiological sources, while inorganic materials do not have ATP. The use of ATP measuring devices, called luminometers, relies on a reaction that occurs between ATP and the chemical luciferase to produce light. The light output, measured in lumens, is measured by the device to indicate the level of ATP present (see Figure 8.2).

The process of using ATP evaluation starts by identifying equipment with a number or a barcode to track and trend performance from ATP swabs. Many devices come with data management software to download data from the device, so you can enter the swab location number and identification into the device. When an ATP test is done, a swab is collected on a visually clean surface and placed into the swab tube. The tube bears a liquid containing luciferin. When luciferin comes in contact with ATP, it is converted to luciferase, which produces light much as a firefly creates light. When the swab tube is placed into the luminometer, it measures the level of light output from the luciferin/luciferase conversion. The light output (lumens) is measured by the meter, and it then provides a readout of the light output, measured in relative light units (RLUs) on the LED display on the face of the meter [6]. This, along with a Pass–Warning–Fail readout, indicates the amount of ATP present on the surface of the equipment being evaluated. Levels for these readout indicators are usually preset by the manufacturer; however, they can be adjusted by the company as sanitation levels improve. A low reading means that there is a very low presence of ATP, which means that there is not a significant amount of organic material present. Since bacteria gain sustenance from organic material, low reading levels are a good indication that bacteria are not present or have no nutrient source on which to grow and that sanitation has done a good job. In this instance, the device will give a "Pass" reading. As levels of ATP get higher, this is an indication that there is more ATP present, thus more opportunity for microbial growth. When the level indicates "Warning," this is an indication that the ATP level has not exceeded the preset "Fail" level, but that it is higher than what is expected and there may be a need to evaluate the cleaning procedure or the equipment for cleanability. Once the ATP level results in a "Fail" reading, the ATP present is high enough to support significant bacterial growth. In this instance, the piece of equipment should be immediately recleaned and the procedure for

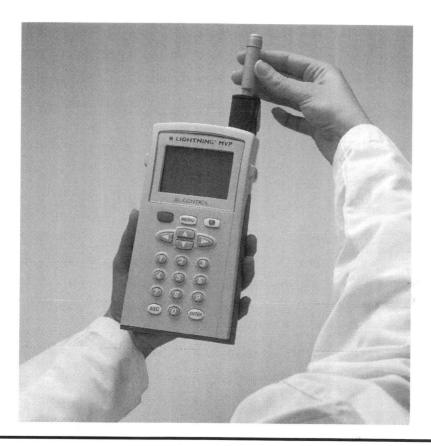

Figure 8.2   The Lightning MVP is an example of a luminometer used to measure the presence of ATP on equipment as a means of further verification of sanitation. (Photo courtesy of Biocontrol.)

cleaning it reviewed for acceptability. Again, the piece of equipment should be evaluated for cleanability and the sanitarian may be observed to ensure there is a complete understanding of the cleaning process and that it is being carried out as defined. What makes this device so convenient is that it gives these results in a minute or less and provides immediate feedback. This is valuable, considering microbiological swabs may take 24–48 hours for results, which means that 1 to 2 days of production may pass on a piece of equipment that is not completely clean, possibly resulting in spoilage or loss of shelf life. The other advantage is that these devices normally store the data collected and are provided with software for data download and management. This means that various pieces of equipment may be tracked for results and the results may indicate cleaning trends.

Does the ATP device indicate the level of bacteria present? Yes and no. Since bacteria is a living organism, it has levels of ATP, so the device may be reading ATP from bacteria, organic soils (meat, fat, etc.), or a combination of both. However, different bacteria have differing levels of ATP. For example, somatic cells have 3000 times the ATP of yeast, while yeast has 10 times the amount of ATP as coliform [12]. So the ATP device is not a good tool to identify levels of bacteria, nor will it identify the type of bacteria present. Therefore, higher results may be from food residue, various bacteria at varying levels or a combination. Lower results, however, are an indication that sanitation has effectively reduced the levels of food soil and bacteria.

The cost of ATP devices varies by several factors: the number of devices purchased or the number of swab tubes used may help in price negotiations. Also, some devices can be purchased with measuring probes for temperature, pH, or conductivity. These probes provide great tools for the sanitation team to measure water temperatures or strength of chemical cleaners and sanitizers. Primarily, they are used to ensure that compounds are at their effective level without exceeding recommended usage rate. This can reduce waste by preventing overuse of cleaners or sanitizers so that they are not above the no-rinse level. They are more accurate than test strips and quicker than titration [1]. The value of the technology is that it provides real time results and requires minimal training to use (see Figure 8.3).

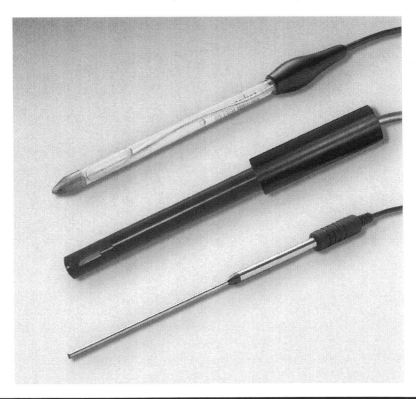

Figure 8.3 Probes that come with the Lightning MVP can be used during the sanitation process to verify the accuracy of preparation of cleaning compounds and sanitizers. (Photo courtesy of Biocontrol.)

The devices usually have a central database when data are logged and downloaded for analysis and continuous improvement action [11].

Microbiological

Even with the use of an ATP luminometer for daily sanitation verification, there is a very real value to microbiological monitoring of equipment for additional verification as well as validation. At regular intervals, swabbing of cleaned equipment and environmental should be performed to confirm the effectiveness of the sanitation program. Any deficiency in certain areas must be corrected immediately and followed by a recheck.

Before beginning a microbiological testing program, a company should ask itself four questions:

1. Why do we want to conduct microbiological testing?
2. What tests or test methods will be used?
3. What locations will be sampled and at what frequency?
4. How will the data be collected, analyzed, and what actions will be taken based on the data?

Why? There are several good reasons as to why a plant will conduct microbiological testing. The primary reason is to gather data to establish verification of sanitation. Taking generic swabs of equipment and the environment provides an objective determination of the efficacy of the cleaning and sanitizing process. Results can be used to identify pieces of equipment that are difficult to clean either due to equipment design, design of the sanitation procedure, or implementation of the procedure. Microbiological counts can be used to help establish cleaning frequency cycles (i.e., every 24 hours, at mid-shift, etc.) for rooms or equipment. Data will also be used for continuous evaluation of improvement of shelf life, product performance, and may result in fewer customer and consumer complaints, less store spoils, and increased sales. Testing and verification data are also employed to support food safety systems [7], thereby avoiding consumer illness, recall, and regulatory action.

What Tests? Determining what types of microbiological testing procedures to use and the methods to follow may be based on the plant, equipment, and products. The outcome may be to involve a combination of tests and materials. Some of the methods available to food manufacturers are described as follows:

Contact plates. These are plates that can be purchased that are already prepared with media. This is an advantage as there is no prep time. They are easy to use with little training. The user simply removes the lid from the plate, presses the plate against the surface to be tested, and places the cover back on the plate. They are incubated according to the manufacturers instructions for the specific agar, and the results are recorded as counts per area. They provide estimates of micro counts since some contaminants won't adhere to the agar [6]. One limitation is that they are best used on flat solid surfaces.

Dipsticks. As with contact plates, these require no prep time as they are already prepared, with media on a flexible stick or paddle. The paddles are in a storage tube attached to a screwtop. Unscrew the top to remove the paddle and place it against the surface to be tested. Return the paddle to the tube and screw the top tight. Incubate according to the manufacturer recommendations and read as a direct count per area [6]. They are easy to use with little training (see Figure 8.4).

Petrifilm or pour plates. Petrifilm is also relatively easy to use and requires less preparation than pour plates. Both can be used for air or, for sample, dilutions

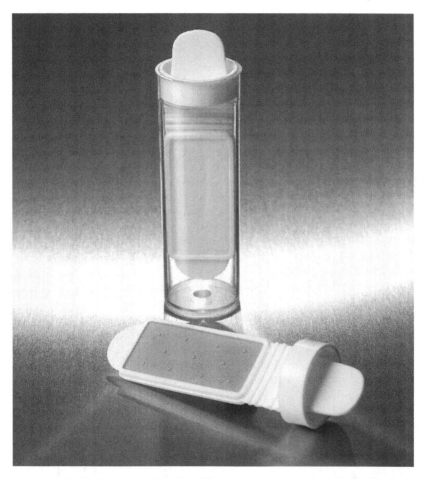

Figure 8.4 Dipsticks are ready to use and provide results for verification of surface cleanliness. (Photo courtesy of Biotrace.)

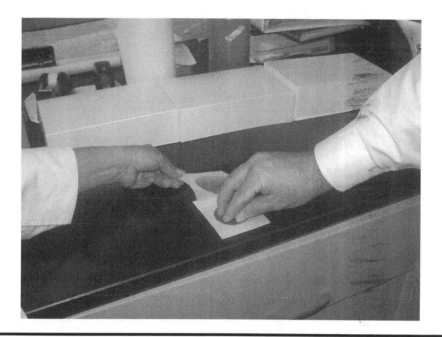

Figure 8.5   Petrifilm may be used as a contact plate for surfaces or for hand contact as verification of hand washing efficacy.

and petrifilm can be used for direct contact with surfaces to verify sanitation. Pour plates or petrifilm may also be used for hand-contact testing as a means of verifying hand washing effectiveness (see Figure 8.5).

When using a sterile swab for sample collection and analysis on petrifilm, it is best to use a template for the area to be swabbed. This provides consistency of the sample area for data analysis. The template must be sanitized before use and between applications to prevent cross-contamination from one site to another (see Figure 8.6).

Protein or carb analysis. These are test strips that can be used to contact production surfaces. Once the strip has been in contact with a cleaned surface, it is exposed to chemicals and there will be a color change if proteins or carbohydrates are present. It is used similar to ATP to detect the presence of these materials from products on product surfaces. Presence of protein or carbohydrate indicates that the surface is either not clean, the protein has been baked on, or a biofilm may be forming.

Rinse test. This is used to measure the entire surface of small equipment parts and involves pouring or soaking the part with rinse water that is collected and analyzed. It is difficult to use with larger parts unless the entire rinse solution can be collected [20]. There may be application of this method with difficult to reach portions of equipment. Again, the entire amount of rinse solution must be collected for analysis.

Figure 8.6 Templates used for microbiological swabbing provide consistency of sample size but should be sanitized between samples.

Air testing methods. Microorganisms may become airborne as passengers on dust or moisture particles. Yeast and mold may be carried by air. Two means of testing are passive and mechanical. When deciding to conduct air testing, consider the location and the climate. The southwest is a dryer climate, thus less conducive to yeast and mold. The southeast with higher humidity has a higher likelihood of mold. Also consider the product that is being made; flour-based processes will have a higher likelihood, whereas meat and poultry processing has a lower likelihood. Use air testing to verify the effectiveness of air filtration or if there are unusual findings in the plant [6] (see Figure 8.7).

Figure 8.7 **Air sampling may be valuable in locations where there is higher humidity or heavy air contamination from food ingredient dust such as flour. (Photo courtesy of Biotrace.)**

Consider the product type made in the plant and the organisms that may be associated with the product when determining the type of testing to be conducted. For example, testing surfaces where product with natural cheese is handled or testing products containing natural cheeses for APC may yield high results because of the cheese culture. This type of test may not produce data that is useful as an indication of quality. However, testing for coliform may be better for product or for environmental swabs as an indicator of cleanliness and sanitary manufacturing. Conducting *Staphylococcus aureus* testing for hand swabs can provide good indicators of efficacy of hand washing.

What Locations? Consider the proximity of the production line to areas that can contribute to insanitary conditions. These areas include trash docks, loading docks, etc. A good sampling plan is required for the plant; otherwise, negative or low results will give the plant a false sense of security [6]. The sampling plan should be science based and have the objective of producing a set of results that represent the conditions that are suitable for analysis. Samples that are not representative of the source are of little use. Also, consider if the samples will be individual swabs, composites, product, or rinse. Too often, a company does not consider the number of samples needed for the information they require, and if the sample size is too small, the chance of detecting a problem is reduced [9]. For statistical relevance, between 5 and 25 samples are required [8]. Test sample size is important; a too-small sample size means that the probability of detecting a problem is small. Determine whether the swabs are to be conducted on a regular basis or whether they are for a specific project and have a beginning and endpoint for collection.

How are data collected and used? Once the samples are analyzed, the data are collected in a format that allows for analysis. This may be in the form of a spreadsheet or graph to demonstrate plant performance. It is not sufficient to simply collect the data without analysis; otherwise, you have not answered the initial question: **Why** collect samples. The data should be used to determine whether cleaning procedures are sufficient, whether extra training of sanitarians is needed, whether plant GMPs are sufficient or in need of revision, or if employees require refresher training. The data may also provide insight into the facility and equipment design for continuous improvement. Data can be used to establish performance of product during shelf life and whether the current posted shelf life can be supported.

Validation

Microbiological evaluation of equipment is an effective means of validating the efficacy of the sanitation process but may require more time than verification. Validation uses data gathered over time to determine whether there is consistency to established processes or procedures.

Environmental *Listeria* Monitoring

Listeria monocytogenes is one of the most significant illness-causing microorganisms for the Ready-To-Eat (RTE) meat and poultry products industry. It is estimated that there are 2500 food-related illnesses per year with almost 500 deaths from listeriosis [20]. Significant outbreaks occurred in 1998–1999 in hot dogs and deli meats, and again in 2002 in turkey deli meats that raised the attention toward the types of products that may be most impacted by *L. monocytogenes* [18]. In 2001, the FDA released a risk ranking of RTE foods associated with listeriosis, and they included deli meats, hot dogs if not reheated, and pâté/meat spreads [21]. Prevention of cross-contamination is the primary means of protecting RTE products, and plants must have very strict Sanitation and GMP requirements to avoid product contamination. However, detection of this ubiquitous organism is also important. It is in the best interest of industry for all RTE meat and poultry product manufacturers to implement aggressive testing programs. The cost of a single *Listeria* outbreak can be staggering to the company responsible, but it also has repercussions across the industry. Eating habits of consumers change, safety rules are reevaluated, and activist groups, acting more on emotion than science, push for tighter regulations. *Listeria* testing for Ready-To-Eat products is a critical and high-profile concern for food manufacturers. Companies have to be vigilant because their product has to be safe when it leaves their facility, and they cannot rely on the consumer to make it safe.

In 2003, FSIS reissued Directive 10,240.4 providing direction to the agency inspection force for the implementation of regulations in 9 CFR 430, Interim Final Rule, Control of *L. monocytogenes* in RTE Meat and Poultry Products dated June 6, 2003. In October 2003, the USDA conducted briefings around the country explaining rationale and expectations for the rule and providing guidance to the industry on preparation of a science-based Environmental *Listeria* Monitoring Program. FSIS expectations are that every plant producing RTE meat and poultry products will have an environmental *Listeria* monitoring program to identify and control microbiological niches that may harbor and support the growth of *L. monocytogenes*. The new rules require manufacturers of RTE meat and poultry products to develop written programs to control *L. monocytogenes* and to verify the effectiveness of those programs through testing. The rule also states that establishments must share testing data and plant-generated information relevant to their controls with FSIS and encourages all establishments to employ additional and more effective *L. monocytogenes* control measures. FSIS ruling results in the addition of direct-contact surface testing for *Listeria* to environmental monitoring programs. Companies have avoided direct-contact testing in the past, waiting to see what the USDA's ruling would be, but the regulations provide some leeway for swabbing the surface and providing *Listeria* spp. or *Listeria*-like results initially. The rule gives manufacturers an opportunity to determine whether contact surface contamination poses a hazard and to deal with it before a problem occurs.

RTE products are either defined by meat and poultry standards of identification, labeled to represent themselves as RTE or expected to be RTE by the consumer. The Rule defines RTE products as follows: "A meat or poultry product that is in a form that is edible without additional preparation to achieve food safety and may receive additional preparation for palatability or aesthetic, epicurean, gastronomic, or culinary purposes. RTE product is not required to bear a safe-handling instruction (as required for non-RTE products by 9 CFR 317.2(l) and 381.125(b)) or other labeling that directs that the product must be cooked or otherwise treated for safety, and can include frozen meat and poultry products." Briefing documents provided by FSIS included a section that gave RTE product examples. The categories of RTE products, with individual examples, listed in Appendix 3 of the briefing manual are identified in Table 8.1.

Non-Ready-To-Eat (NRTE) meat and poultry products are not covered by this rule. These are products that may contain a RTE component but by their nature

Table 8.1 RTE Meat and Poultry Products

Category	Products
Deli meats	Bologna
	Cappicola
	Chicken bologna
	Cotto salami
	Lebanon bologna
	Pepperoni
	Roast beef
Dinners	Dinners have at least
Entrees	Burritos
	Chicken breast
	Chili
	Gyros
	Meat loaf
	Pasta with meat sauce
	Ravioli
	Stews
Hot dog products	Chicken franks
	Turkey franks
	Weiners
Other nonsliced sausage	
Snacks/hors d'oeuvre	Beef sticks
	Carne seca
	Meat/poultry jerky
Thermally processed, commercially sterile products	Canned spaghetti and meatballs
	Canned chicken salad

Source: U.S. Department of Agriculture, *Listeria monocytogenes* Workshop, Washington, D.C., 2003.

would not be eaten as is by the consumer. They may include a component that requires cooking to make the product safe, for example, fully cooked meat patties with an uncooked batter/breading coating. These types of products must bear labeling that clearly identifies to the consumer that the product must be fully cooked prior to consumption. Example of this type of labeling includes "Ready To Cook" and "Cook and Serve." Cooking directions on the package must also be validated to ensure that, if followed, they will provide adequate lethality for product safety. Finally, the label must include the Safe Handling Statement.

Alternatives

Plants producing RTE products and implementing a required program will also identify the process by which their products will be maintained safe for consumption. These processes fall into one of three alternatives.

1. **Alternative 1:** Products in this alternative apply both a post-lethality treatment and an antimicrobial agent or process to control the growth of *L. monocytogenes*. A post-lethality treatment is applied to the finished product or product sealed in a package. Examples of post-lethality treatment are steam or hot water pasteurization, high hydrostatic pressure treatment, or use of ozone (i.e., on franks after peeling). These treatments may be dependent on factors such as temperature or exposure time. Antimicrobial agents are used in the product formula to reduce, eliminate, or suppress the growth of *L. monocytogenes* through the shelf life of the product. Two of the more common antimicrobial ingredients are potassium lactate and sodium diacetate. An antimicrobial process such as freezing may not kill the organism but will suppress growth as *Listeria* does not grow below 31°F. All treatments and processes must be validated for effectiveness through testing or scientific

literature. Post-lethality treatment must be included in the HACCP plan as a CCP and the antimicrobial agent or process must be included in the HACCP plan, SSOP, or a prerequisite plan. FSIS has indicated that products falling into Alternative 1 will receive the least amount of regulatory testing as they are considered to be at a lower risk than Alternatives 2 or 3.

2. **Alternative 2:** Products in this alternative include either a post-lethality treatment similar to that described in Alternative 1 or an antimicrobial agent or process described in Alternative 1. The post-lethality treatment must be included in the HACCP plan and must be validated for its effectiveness in killing *L. monocytogenes*. When an antimicrobial agent is used, it must be included in the HACCP plan, SSOP, or prerequisite program to document that the agent suppresses growth. If only an antimicrobial agent or process is used, then the plant must maintain effective sanitation because antimicrobial agents are not as effective at high levels of contamination [22]. The plant must also include testing program for food-contact surfaces for *L. monocytogenes*, *Listeria* species, or *Listeria*-like organisms in the post-lethality environment. The testing program must identify the swab site locations and swab size, the frequency of swabbing, and justification as to why the frequency is sufficient and procedures for holding and testing product in the event of a positive *L. monocytogenes* on a direct product contact surface. FSIS has indicated that products falling into Alternative 2 will receive less regulatory testing than Alternative 3.

3. **Alternative 3:** Products in this alternative do not incorporate any of the previously identified intervention strategies and rely on sanitation for the prevention of post-lethality contamination of product. Because *Listeria* is continually

Testing for *Listeria*-like organisms refers to the process of identification of organisms found in the environment as Gram-positive bacilli. This involves collecting a swab and enriching it in Fraser broth for 24–48 hours. If during this time the broth turns dark, it is Gram-stained for further identification. A result of Gram-positive bacilli may or may not be due to a *Listeria* organism as there are other bacteria that fall into this category such as staph and bacillus. However, it is an indication that the environment provides conditions that will support the presence and growth of *Listeria*. The advantages of *Listeria*-like results is that it takes less time for identification than to species, meaning that corrective and preventive actions can be implemented sooner and it may be more cost-effective [17].

Testing for *Listeria* species indicates the possible presence of all strains and also means that the environment can support the presence and growth of *L. monocytogenes* but does not mean that the organism is present [13]. Both findings indicate that corrective actions are needed.

introduced to the food environment and can reestablish and then be transported through the environment, it is vital that interventions be implemented. FSIS considers multiple interventions more effective than single interventions [22]; therefore, it considers products in Alternative 3 to be at the highest level of risk. Thus, direct product contact testing is required in the SSOP for these products. As with Alternative 2, the testing program must identify the swab site locations and swab size, the frequency of swabbing and justification as to why the frequency is sufficient, and procedures for holding and testing product in the event of a positive *L. monocytogenes* on a direct product contact surface. A positive for *Listeria species* or *Listeria*-like organisms will require corrective actions and testing verification that the actions were effective. A second positive result on a follow-up test will require additional corrective action and a test, and hold procedure for production lots until the deviation is corrected. Lots in the test in hold may be released if they are sampled with a program that will show a 95% statistical confidence. Any positive for *L. monocytogenes* on a direct product contact surface results in the product being considered adulterated and subject to rejection. FSIS has indicated that products falling into Alternative 3 will receive the most regulatory testing.

For all three alternatives, FSIS expects the establishment to maintain sanitation in the post-lethality environment. However, FSIS assumes that sanitation is only 85% effective at controlling *Listeria* in the environment, so the plant must demonstrate that their sanitation program is more than 85% effective [22]. If *L. monocytogenes* control measures are included in the HACCP plan or plant SSOPs, the plant should validate and verify the effectiveness of the measure. If a plant determines that *L. monocytogenes* is not a hazard reasonably likely to occur in the post-lethality environment, and thus not a Critical Control Point in the HACCP plan, then they must have a prerequisite microbiological testing program to detect niches, or places where the organism can hide, live, and multiply. If the plant opts for a prerequisite program, they must maintain documented results and make the program and results available to agency personnel.

Prerequisite Monitoring Program

The plant will develop a "science-based" Environmental *Listeria* Monitoring Program to monitor the effectiveness of sanitation and other preventive programs and confirm that control measures are sufficient to maintain a sanitary environment in the RTE production area where finished product is stored, processed, or packaged [18]. The program will meet the requirements outlined in the Final Rule and should include the following information: Identification of Post-lethality Swab Locations, Means of Swab Site Selection, Collection of Swabs, Analysis of Swabs, Data Collection and Management, Corrective Actions, and Test and Hold Scenarios [13]. The company goal should have zero tolerance for *Listeria* in the environment.

This goal may not be achievable based on the ubiquitous nature of the organism; however, it is an aggressive goal that can help reduce the frequency of findings in the environment that can translate to minimal findings in finished product.

Identification of Swab Sites

Identification of swab sites involves a logical approach that includes understanding the processing plant's history with *Listeria*, and an objective to seek locations with a high probability of identifying areas where the organism will harbor and thrive. Because *Listeria* can be difficult to find, the program must be aggressive with the goal of testing to find positive locations [25]. If the testing program is effective at identifying these harborage sites and finding *Listeria*, action can be taken to eliminate them. Over the years the plant can build and hone an aggressive environmental monitoring program that can reduce the presence of *Listeria*. To organize the testing program, a cross-functional team that includes the plant QA Manager, production, maintenance, and sanitation will break the facility down into three categories: direct product contact sites, indirect contact sites, and noncontact sites. For purposes of regulatory requirements, the program only requires testing of direct product contact surfaces, but to be effective, the entire plant should be swabbed to identify the harborage locations that can eventually contribute to contamination of direct product contact surfaces. Each category is then broken into separate target sites, which are then numbered for identification. Examples of the target sites in each category are direct contact surfaces, such as conveyor belts, fill hoppers, and scoops that come in direct contact with the product; indirect contact surfaces, such as table legs, control panels, or broom handles that have proximity to products or might be used by someone who incidentally handles the products; and noncontact surfaces, which include walls, floors, drains, overhead pipes, or any surface that does not have contact with products but may harbor pathogens.

Should drains be swabbed? Yes. Drains usually are harborage sites for *Listeria* because they are damp and contain sufficient nutrient source. This would not be a problem if the moisture in the drain stayed in the drain; however, this is not always the case. Drains can back up resulting in contaminated material forced into the plant that can then be transferred to contact surfaces. At times during the sanitation process, the drain can be hit with high-pressure water that creates an aerosol mist that can contaminate lines. For these reasons there should be some level of testing the drains, even if it is less frequent than other sites (see Figure 8.8).

Commando Swabs

When setting up the initial program, a plant may not have a sufficient amount of data if the plant is new or has not previously conducted an environmental program. One means of developing a significant number of samples in a short amount of time is to conduct what is referred to in the industry as commando swabs. The writer

Figure 8.8   Drains present a particular challenge and should be part of the swabbing process.

is not certain of the origin of the term, but it fits with Webster's definition of *commando* as a small raiding party operating within enemy territory. If the plant is the territory and *L. monocytogenes* is the enemy, then a small group of people within the plant will conduct commando swabs.

The group will evaluate the flow of product, equipment, and people to begin with. They will use this evaluation to establish where in the plant there is exposed in-process or finished RTE product and to identify areas where there can be cross-contamination. To some extent, there are no limits placed on the locations where the swabs are collected to get a broad view of conditions in the plant. With the exception of the raw product handling areas, there are no areas of the plant that cannot be swabbed. Sites will be identified with a number and corresponding description prior to swab collection. Depending on the plant size, this method may involve approximately 100 swabs per day over a two- to three-day period. Each day the same locations will be swabbed and the swabs analyzed *Listeria-species*. Testing to *Listeria species*, as opposed to *Listeria*-like organisms, provides the plant with a much clearer idea of the presence of *Listeria* in the environment. Once the results are received, they will be evaluated to identify repeated positive areas or potential trouble spots and the sites for the routine sampling program can be identified. The writer likes to use this method in plants where there is a positive product or contact surface finding as it provides a broad investigative tool. It is also effective to use on an annual basis as plants or processes change to determine whether there are new or evolving niche areas.

Swab Frequency and Selection of Sites

Once the sites have been identified, they will be scheduled for testing on the frequency identified by the plant. The rule recommends the following frequencies for each alternative:

- Alternative 1: At least twice per year due to the use of multiple implementation strategies.
- Alternative 2: At least four times per year because of the use of only one intervention.
- Alternative 3: Monthly if the plant produces non-deli or non–hot dog type of products, four times per month for large volume deli or hot dog producers, or twice per month for small volume deli or hot dog producers.

The frequency of testing may also be a function of the processing environment. For example, manufacturing in wet areas may be tested more frequently because wet areas support *Listeria* growth better than dry areas, until a baseline can be established. The number of samples collected and the frequency may be adjusted based on the results over time. Frequent negative results at a swab site may support elimination of that site from swabbing in favor or another location that may be suspected as harborage locations. Swabs may be composited when scientifically appropriate; however, no more than five samples should be composited [17]. Whichever frequency a plant determines that they will use, the plan will include an explanation of why the testing frequency is sufficient to ensure the effective control of *L. monocytogenes* or the indicator organism. Each site should have an equal opportunity for selection. One means of ensuring the randomness of site selection is through the use of an Excel spreadsheet to select random numbers. Once the site locations are organized by category, the site numbers are placed in a table on the spreadsheet, and when the formula is input, it will generate random number selections using the = RANDBETWEEN(N,N) formula. The "N" represents the site numbers previously identified by the cross-functional team, and the portion in the brackets represents the range of site numbers identified for the plant. This means the plant does not have to pull the numbers out of a hat to be sure they are random.

An example of the random site generator is illustrated in Table 8.2. The spreadsheet can also be programmed to randomly select the day of the week and the shift when samples are to be collected.

Collection of Swabs

The spreadsheet can be programmed to identify when during the day the sampling will be done: either at pre-op or during operations. The agency prefers that samples are collected during production operations; however, there is great value to collecting the samples at pre-op to verify and validate the effectiveness of cleaning.

Table 8.2   Random Site Generation

Cleo's Foods Environmental Monitoring Program Random Site Selection Generator			
Zone	**Site**	**Week of:** 10/29/2012	
1	10		
	6	**Day**	**Pre-op/op**
	6	5	2
	17		
	19	**Key**	**Key**
2	44	1 = Monday	1 = Pre-op
	34	2 = Tuesday	2 = Op
	26	3 = Wednesday	
	26	4 = Thursday	
3	53	5 = Friday	
	47		
4	66		

Zone key	**Examples**
1 = Direct contact	(conveyor belts, hoppers, filling tubes, scoops, tables)
2 = Indirect contact	(frames, drive rollers, utensil handles)
3 = Noncontact	(floors, walls, wheels)
4 = Auxiliary	(welfare room floors, raw product areas, corridors)

The writer suggests that the plant should not assume that cleaning has been totally effective at eliminating *Listeria* organisms and collect swabs at pre-op to validate the sanitation process for a period of time. However, once the plant has sufficient data to verify that sanitation is effective, it will be preferable to focus their swabbing during operations. When swab samples are collected during operation, the equipment should be running for at least 3 hours as this will make a hidden source more detectable [24]. As equipment operates, there is more opportunity for liquids or solid soils that is harbored in niches, such as hollow rollers, bearings, or overlapping equipment, to loosen and work its way onto product surfaces or product.

Sites can be swabbed with a sterile sponge or sterile gauze; however, Q-tips are usually not recommended as they can be fragile and limit the amount of material that can be picked up as compared to sponges or gauze. Q-tip type swabs are helpful for small recesses or holes into which a sponge cannot reach. The swab material should be premoistened with neutralizing buffer, such as Letheen broth, to limit the effect of sanitizer used in the plant [18]. This can be done at the time the sample is being collected or sterile sponges can be purchased in resealable sample collection bags that include the neutralizer. They may also be purchased with a separate attached bag with sterile gloves. It is highly recommended that prior to collection the person conducting the swabbing will receive documented training on swab collection techniques. This will include donning a clean smock and hairnet as well as washing and sanitizing their hands to prevent unintended contamination of the swab.

Sample bags will be identified with the date, plant ID, site number, and time of sample collection. This should be done with an indelible marker, so the information does not rub off the bag. With clean and sanitized hands, grasp the sample bag in one hand and push the sponge to the top of the bag with the other hand. Open the top of the bag, then put on sterile gloves. Grasp the sponge with a gloved hand and swab the sample site. The area swabbed will be approximately 1 foot by 1 foot depending on the size of the swab site. However, do not limit your swab area to 1 foot by 1 foot; if more area is available, swab as much of the area as possible. Rub the sponge across the area vertically, horizontally, and diagonally to cover the entire surface. Use hard pressure when collecting the swab to ensure that any organisms that may be in small niches or imperfections in the surface are picked up. As previously stated, the objective of swabbing is to find *Listeria* or *Listeria* harborage. Do not limit your program to simply meeting minimal regulatory or customer requirements [26] (see Figure 8.9).

Return the sponge to the sample bag, and securely seal the bag. Dispose of the gloves prior to collecting the next sample.

Once the swabs have been collected, they will be refrigerated until they can be analyzed or shipped to the designated third-party laboratory for analysis. If they are being shipped to a third-party lab, they will be placed in a Styrofoam shipping container and secured with newspaper or other packing material. Include a sufficient number of frozen ice packs to keep the samples adequately refrigerated. If dry ice is used, make sure that it does not come in direct contact with the samples as this may damage the sample bag and render the sample useless. Include a Lab Sample Submission Form identifying the company name, sample identification numbers, date of collection, and testing required. An example of a sample submission form can be found in Table 8.3. Samples should be shipped the same day as they are collected but no later than the next production day. Airfreight shipments should be for overnight delivery, and the laboratory is to be directed to begin setting the samples on the same day of receipt, even if this is a weekend day.

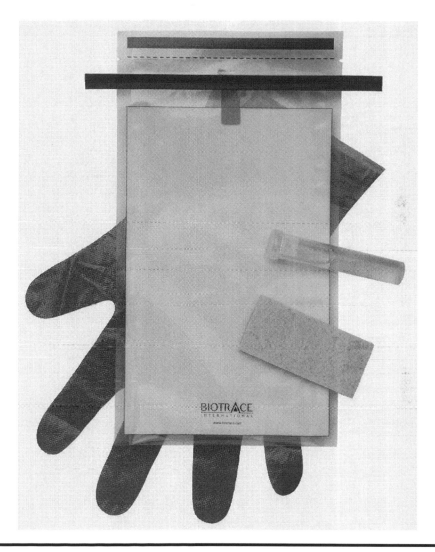

Figure 8.9 Using a sterile sponge with sterile gloves and neutralizing broth for the collection of an environmental swab sample. (Photo courtesy of Biotrace.)

Table 8.3 Lab Sample Submission Form

LABORATORY SAMPLE SUBMISSION FORM **Client Information**				
Company Name: _____ Cleo's Foods _____				
Contact Name: _____ Cleo Katt _____				
Address: _____				
City, State, Zip: _____				
Phone Number: _____				
Fax Number: _____				
Submit to (Check One):				
Food Testing Laboratory Services, Ltd.				
Purchase Order Number: _____ Date: _____				
Sample Identification and Description	*Analysis to Be Performed*	*Method of Analysis*	*Product Specification*	*Special Instructions*

Analysis

Whether testing in the plant or at a third-party testing laboratory, the method used will be the current USDA/FSIS targeted method for analysis of samples, and this reference should be identified in the plant program. The plant may choose to have the initial swab samples analyzed only to Fraser results, and this testing should be referred to as "*Listeria*-like" for the environmental samples. Negative Fraser results indicate the area represented by the swab is clean and free of *Listeria*. Fraser positives are Gram-stained, and those that are Gram-positive bacilli indicate that conditions would support the growth of *Listeria*. As previously indicated, *Listeria*-like findings do not mean that *Listeria* is present, but they do not rule out the possibility of the presence of a *Listeria* species, including *monocytogenes*. The advantage of *Listeria*-like is that the result comes to the plant faster, and this means that the plant may act on the result quicker.

The plant may also choose to have the results carried out to identify that the organism is a *Listeria* species. This analysis is preferred by USDA and, if positive, will demonstrate the presence of a *Listeria* organism, but does not confirm whether it is pathogenic *monocytogenes* or one of the other nonpathogenic strains. Lastly, the plant may opt to have the test carried out to species identification; however, they must keep in mind that any finding of *L. monocytogenes* on a direct product contact surface means that the product run across that surface is considered by USDA to be adulterated and subject to disposal.

PATHOGEN TESTING: IN PLANT OR OUTSOURCE?

Plants conducting environmental *Listeria* monitoring have a choice of several testing methods that can be employed in-house. These include the use of pour plate, Petrifilm, and lateral flow devices or more complex equipment using Polymerase Chain Reaction (PCR) technology. As part of Good Lab Practices, the laboratory should maintain positive controls to verify the accuracy of their test results. The dilemma that exists with in-plant testing is the potential for contamination from laboratory samples or positive controls being transferred to the manufacturing plant. If the testing lab is physically separate from the plant, there is a margin of safety. However, if the laboratory is attached to or inside the manufacturing facility, there is reason for concern and strict control measures must be taken to prevent contamination from the lab being transferred into the plant. Though it may be more expensive, the writer recommends that pathogen testing be conducted at a third-party lab that has A2LA or ISO certification.

Data Tracking

Using Excel spreadsheets, a plant can develop a simple yet comprehensive data tracking system for every *Listeria* test result that occurs inside the plants. The spreadsheets identify troublespots early on, track percentages of positive Fraser results, and most important, create documentation verifying that potential hazards have been dealt with and corrected. With a spreadsheet program such as Excel, you can log data results, calculate the number and percentage of positive tests, and document your actions taken, all in one document (Table 8.4).

Fraser positive results are noted as "ones" in the spreadsheet. When a Fraser positive shows up, the swab area is identified, recleaned, and examined for possible damage or harbor areas. If the positive is in an indirect or direct contact zone, after being cleaned and sanitized, the area is retested three times a day for three days— before the morning shifts begin, 3 to 4 hours into first shifts and 3 to 4 hours into the second. A program will identify the number of Fraser negatives required to prove the problem area has been dealt with effectively. If the plant chooses to test to

Table 8.4 Environmental Monitoring Spreadsheet

Cleo's Foods Environmental Monitoring Program			Date	10/1/03	10/9/05
Zone Key:			**Pre-op or Op**	Op	Op
1 = Direct contact					
2 = Indirect contact					
3 = Noncontact					
Result Key:					
0 = Negative					
1 = Fraser positive					
2 = *L.* spp. positive					
Test Site	Site Name	Zone	% Fraser	% L. spp.	Retest Y/N		
1	Peeler horn	1	0.00%	0.00%		0	0
2	Product conveyor	1	0.00%	0.00%		0	0
3	Conveyor frame	2	50.00%	0.00%		1	0
4	Peeler housing	2	50.00%	50.00%		0	2
5	Floor	3	0.00%	0.00%		0	0
6	Drain #3	3	0.00%	0.00%		0	0

Listeria species as opposed to *Listeria*-like, and confirm the species, then "1" will be used to represent *Listeria species* and "2" will represent *L. monocytogenes.*

The spreadsheets provide the company with enough data to see at a glance which sites might have trouble areas or require extra attention and which areas show a history of troubles, enabling him to target potential hazards long before they get out of control.

These results, combined with daily organoleptic evaluations and equipment testing using ATP bioluminescence technology to identify presence of organic material, also provide information on the sources of environmental contamination, the extent to which contamination is present, and provide information on equipment design and sanitary operations in post-lethality areas. This cumulative information allows the plant to take corrective action to eliminate potential growth spots. If repeated positives appear, the equipment is examined for scratches or crevices that make it difficult to clean properly. The daily testing helps identify problem areas before bacteria growth can occur. Use the data to validate the efficacy of the sanitation program.

Positive Swab Investigation

Once a *Listeria* positive swab result is received, it is very important that the plant identify the source of the organism. This is accomplished by conducting a thorough investigation of the positive site. The investigation will be conducted to identify the possible sources of the positive result, the corrective action (s) that need to be taken, and preventive measures to avoid future contamination. Investigations should take place as soon as possible following receipt of the positive result, preferably the day following confirmed positive results, and no later than 48 hours after receiving results. The steps of the investigative process are as follows:

1. Assemble the Team: The investigative team will be cross-functional and will include, at a minimum, QA, Maintenance, Production, and Sanitation. Participation by Maintenance is very important as they will be instrumental in taking apart equipment or opening up panels to enhance the depth of the investigation. Designate one team member to be the "scribe" who will document all observations and one to be the photographer to take pictures of key findings. Finally, designate a person who will collect swabs. This team member must have received documented swab training.

2. Materials Needed: During the investigation, it will be necessary to take photos of findings requiring corrective action, so have a digital camera available. Use a form for documenting findings such as equipment design that creates harborage or employee practices that can lead to contamination. Bring 15–20 sponge swabs so that swab samples can be collected from suspect sites.

3. Conduct the Investigation: QA will be the team leader, so they will inform the team of the specific site that tested positive. The team will go to the site and begin the investigation. Effective investigation will include the following:

    a. Look for areas that are difficult to clean due to poor sanitary design, damage, or that have soils present. Also look for potential harborage in the environment such as exposed wood or insulation on freezers, low floor areas where water can pool, overhead condensation, or cracks in walls.

    b. Observe employee practices in the area that may be nonconforming with plant GMPs. Look for personnel who are coming from raw product areas, personnel who are handling wood pallets or materials on the floor and then coming in contact with lines or products, or those who are not thoroughly washing their hands.

    c. Collect a swab from the original positive site as well as from any site that the team members identify as suspect for *Listeria* harborage. There is no set number of swabs that should be collected for an investigation, but a recommendation will be no less than 3 up to approximately 10 depending on the complexity of the equipment and surrounding area. With any luck, the swabbing will identify a positive site that can be deemed the source of the original positive result and definitive corrective action that will eliminate the source.

    d. Document all findings that may need correction. The team will not be constrained by cost or complexity of the correction. This is the opportunity to record all findings they believe could have resulted in the positive finding. The listing of recommended corrective actions will be reviewed by the team with the appropriate plant management following the completion of the investigation. During this time, the management team will determine the cost, complexity, and priority of corrective actions.

4. Take Corrective Actions: Once there has been a determination of the priority of actions, take the actions that can be taken immediately at the end of production that day. Other actions can be taken over time as capital dollars, contractor time, or materials are available.

5. Conduct Intensified Sanitation: This will be conducted at the end of production following completion of the investigation. This may require breaking down equipment to access potential harborage sites, scrubbing of broken down equipment parts, application of high-level sanitizer, or application of a different sanitizer.

6. Collect Intensified Swabs: The day following the investigation QA will begin collecting intensified swabs from the original positive site. This means collecting swabs three times a day for three consecutive days. Swabs will be collected at preop and twice operationally after the line has run for at least 3 hours. The purpose of intensified swabbing is to demonstrate that the initial corrective actions and intensified sanitation were effective at eliminating the source of the positive result. Collection of the swabs demonstrates repeatability of the negative results. However, if any of the intensified swabs tests positive, then it will be necessary to conduct additional investigation, further corrective actions, and more intensified cleaning and swabbing until the source can be confirmed to be eliminated.

Corrective Actions

The corrective actions should specify the cleaning procedures of equipment with positive direct food contact results. This includes the dismantling of equipment. More intensive cleaning procedures or activities should be described for repetitive positives. Corrective actions should not only include the retesting of positives but also the expansion of the testing areas and the inclusion of more site-specific testing in the immediate area of the positive. Corrective actions for the indirect contact areas should include the general area cleaning to keep the surrounding product and nonproduct surfaces from being contaminated. Review and repair of any deficiencies in equipment or facilities both should be considered. Corrective actions should include the investigation of the causes and the program changes that may be required to keep a repetitive condition from being established. This should specify a review of prerequisites other than sanitation items such as employee hygiene, the maintenance program, and product handling (Table 8.5).

All of the corrective actions, from the cleaning to the repeated swabs, can be noted in a spreadsheet's comment tab. This is an important part of the process because it is not enough to have volumes of data that point to problem areas; there must be documentation showing how the problem was addressed, and proof that corrective actions were effective.

As an example, if a company tests to *Listeria*-like, then Fraser's positive findings will require corrective/preventive action and retesting based on the location and proximity to product. Noncontact location Fraser-positive results may require investigation as to the cause of the positive result. This may include, but not be limited to, review of cleaning and sanitizing procedures, intensified cleaning action, observations of sanitation personnel to verify procedures are adequate, and observation of physical conditions and repair of findings of damage resulting in organism harborage.

Indirect contact site Fraser-positive results may require further investigation as to the cause of the positive result. This may include, but not be limited to, review of cleaning and sanitizing procedures, intensified cleaning action, observations of sanitation personnel to verify procedures are adequate, and observation of physical conditions and repair of findings of damage resulting in organism harborage. In this case, follow-up swabs of positive indirect contact locations will be collected on the day following corrective action to provide verification that the action was effective. Follow-up swabs can be collected at pre-op, during first shift operation, and during second shift. If all swabs are negative, the site location will return to normal (random) testing frequency. Positive swabs may require that additional corrective action will be taken and the area swabbed again. If these swabs are Fraser positive, they will be carried out to determine whether they are *Listeria* species positive. If the reswabs are negative for *Listeria* spp., the sites will return to normal (random) testing frequency. If sites are positive for *Listeria* spp., further corrective action is to be taken to identify and eliminate the source of the contamination and the site will be reswabbed. If the reswab is negative for *Listeria* spp., the site will return to normal

Table 8.5 Corrective Action Plan

Cleo's Foods Environmental Monitoring Program
Corrective Action Plan

Date	Site Number ID	Result	Corrective Action	Action Date	Responsibility	Effectiveness
10/1/2005	3	Fraser +	Frame evaluated for pits or damage, recleaned and sanitized at higher concentration of quat (800 ppm).	10/4/2005	Sanitation	Follow-up swabs tested negative.
10/9/2005	4	*L.* spp. +	Peeler evaluated for pits or damage, recleaned and sanitized at higher concentration of quat after baking in the smokehouse.	10/14/2005	Maintenance	Follow-up swabs tested negative.

(random) testing frequency; however, if it is positive for *Listeria* spp., the test will be continued to identify the species. If the test confirms the presence of *L. monocytogenes*, the equipment will be taken out of service until the source of contamination is identified or eliminated. Once the source is identified and eliminated, the equipment will return to normal (random) testing. Additional investigative swabs, outside the parameters of the positive test site, may be taken to aid identification of source and cause. These steps are not required by the rule; however, they represent an aggressive program geared toward preventing the organism from becoming established in the environment and eventually contaminating contact surfaces.

Because of the implication to product, direct product contact Fraser-positive results will require the greatest amount of investigation as to the cause of the positive result. This should include, but not be limited to, review of cleaning and sanitizing procedures, intensified cleaning action, observations of sanitation personnel to verify procedures are adequate, and observation of physical conditions and repair of findings of damage resulting in organism harborage as well as dismantling of equipment for aggressive cleaning. Follow-up swabs of positive contact sites should be collected the day following corrective action to verify effectiveness of corrective actions. Swabs may be collected at pre-op to verify effectiveness of sanitation and during first shift or second shift operations if moving equipment is suspected as the source. All swab results must be negative to resume normal sampling protocol.

If any results are positive for *L. monocytogenes* during the retesting phase, the equipment should be removed from service for thorough evaluation until the source of the positive is identified and eliminated. Taking equipment out of service is an aggressive approach because the line cannot run. It is better the line does not run than to send out suspect or contaminated products. Once the equipment is returned to service, collect swabs to continue to verify that corrective actions were effective. If the program dictates that product sampling is required, the product from the line tested is placed on QA Hold, from cleanup to cleanup on the line tested, until acceptable results are received. All of these additional actions are documented in the spreadsheet as well.

The program must clearly state that results produced by implementation will be maintained in the documentation required under 9CFR417.5 and that the results will be available on request to FSIS inspection personnel.

Test and Hold

To further comply with 430.4, Control of *L. monocytogenes* in Post-Lethality Exposed RTE Products, if the plant employs only an antimicrobial process or agent to suppress the growth of *L. monocytogenes,* then the plant will test food contact surfaces in the post-lethality environment and the program must also include identification of the conditions under which the establishment will implement test-and-hold procedures following a positive test of a food contact surface for *L. monocytogenes* or an indicator organism. As an example of this circumstance, if the initial contact surface test result is positive for *L. monocytogenes* or an indicator organism, the plant will take corrective action and conduct targeted corrective action of the specific site and additional surrounding sites they deem necessary to ensure effectiveness of the corrective actions. If there is a second positive test result at the targeted site, the plant should implement a test-and-hold procedure until the site is clear [22].

Product made on the line where the site tested positive must be tested for *L. monocytogenes* or an indicator organism using a sampling method that will provide sufficient statistical confidence that the lot is not adulterated with *L. monocytogenes*. In the Compliance Guidelines to Control *L. monocytogenes* in Post-Lethality Exposed Ready-To-Eat Meat and Poultry Products, FSIS refers to the International Committee for Microbiological Standards for Foods (ICMSF) sampling plans as dictated by cases [22]:

- Case 13 conditions reduce the hazard (i.e., product will be cooked or contains an inhibitor). Collect 15 product samples for analysis. Release if there are 0 positive test results.
- Case 14 conditions result in no change to the hazard (i.e., frozen or shelf stable products). Collect 30 samples for analysis. Release if there are 0 positive results.
- Case 15 conditions may increase the hazard (refrigerated product supports *L. monocytogenes* growth). Collect 60 samples for analysis. Release if there are 0 positive results [22].

Special Event Sampling

Though this is not identified as a requirement in the rule, plants should consider the potential hazard created by dust and traffic during construction and include this testing in their protocol. During periods of construction, aggressive sampling should be conducted as determined by the plant QA manager and plant manager with maintenance input. The normal (random) line sampling should continue during this period with additional environmental samples collected in addition in the vicinity of the construction. Sampling of floors, walls, and drains (noncontact), as well as indirect sites adjacent to the construction area will be conducted on commencement of construction. *Listeria*-like or *Listeria* spp. positive swabs will require documented corrective action and immediate re-swabbing to verify effectiveness. Repeated abnormal findings will result in additional cleaning and sanitizing, reinforced GMP training for contractors and crews and plant employees.

FSIS Testing

In Directive 10240.3, FSIS identified four types of testing it will conduct for deli types of products [24]:

1. Intensified: FSIS will collect multiple samples of contact and indirect contact surfaces and will conduct increased record verification.
2. Targeted: FSIS will randomly select one product at a time for testing but will not collect environmental swabs or conduct record review.
3. Low-targeted: FSIS will sample product at a lesser frequency than targeted testing.
4. Nontargeted: FSIS will sample "as necessary."

If USDA collects product samples for any pathogens (e.g., *Listeria, Salmonella, Escherichia coli* O157:H7, etc.), all production from the line sampled (cleanup to cleanup) will be placed on QA Hold pending acceptable (negative) results. Verify that product is placed on QA Hold at distribution. Preshipment paperwork will be signed when all required CCP monitoring and verification is completed. Request the regulatory agent to provide you with a sample log number and a copy of the government lab results as quickly as possible. Do not split samples or analyze duplicate samples. In the event of a pathogen-positive result, from the regulatory agency, product will be disposed as soon as possible in a sanitary landfill.

Intensified Verification Testing and Routine L. monocytogenes *Monitoring*

For the last several years, FSIS has implemented practices whereby they collect swabs from the post-lethality production environment as well as finished RTE product for *L. monocytogenes* analysis. The collection falls under one of two scenarios, Intensified Verification Testing (IVT) or Routine Risk-Based *L. monocytogenes* Sampling (RLm). These are typically conducted by an Enforcement, Investigation, and Analysis Officer (EIAO). The IVT is often conducted as the result of a prior *Lm*-positive finding on a product contact surface or during product testing. The RLm is often in conjunction with a Food Safety Assessment as directed by the district office [27].

When FSIS collects swabs and product samples during an IVT or an RLm, all production from the line(s) sampled (cleanup to cleanup) will be placed on QA Hold pending acceptable (negative) results. If product has to be held in a third-party warehouse due to plant space constraints, verify that product is placed on QA Hold at the distribution. Do not release product or ship it to customers until negative results are received from FSIS. In addition, FSIS personnel will be observing plant practices when they are conducting an RLm or IVT. If the plant significantly alters its routine practices, the EIAO may choose to postpone the sampling until a later time and may issue an NR to the plant. During either sampling process, it is best that the plant maintain normal cleaning and production practices. It is acceptable to limit production by shortening the day; however, there must be enough production for the EIAO to collect swabs at pre-op, during production, and toward the end of production [28].

FSIS views positive *L. monocytogenes* results on a direct contact surface as indication that product produced on that surface is adulterated. If a direct product contact surface swab is positive for *L. monocytogenes* or if product tests positive for the pathogen, all product from the tested line(s) will be destroyed by disposal in a sanitary landfill. It is also likely that FSIS will take some type of regulatory control action, such as issuance of Notice of Suspension. In this instance, the plant will have to demonstrate through investigation and follow-up swabbing that the environment is free of *L. monocytogenes* and that product is not subject to adulteration. Once this is provided to FSIS, then the USDA will reinstate inspection.

Positive results from indirect or noncontact surfaces will require follow-up corrective action to prevent further incidents of positive results or possible translocation to product surfaces or product. In these instances, it is less likely that FSIS will take regulatory control action. However, they may require that the plant demonstrate through investigation, corrective action, and intensified swabbing that the harborage has been eliminated. It is also likely that they will conduct a follow-up IVT or RLm in the near future to verify that the plant actions have been effective.

Product Testing for L. monocytogenes

The question of whether to collect product samples for *L. monocytogenes* analysis is a subject of debate. According to Henning and Cutter, it is one of the best methods to verify the environment and safety of the product. However, because only a small sample size is collected, the organism may not be detected and the result may provide a false sense of security about product safety [13]. Since *Listeria* is not uniformly distributed in product and not found frequently there, product testing may not be a reliable indicator that *L. monocytogenes* contamination has not occurred [17]. Product testing will not indicate the mode or source of contamination if there is a positive result, whereas environmental testing will indicate potential sources that can be targeted for corrective action [24].

If product samples are collected, they must be protected from incidental sampling contamination. Product packaged for retail sale (i.e., 8 oz. sliced vacuum-packaged deli meat, 1 lb vacuum-packed hot dogs) will be collected intact in the original package for shipment to the lab. Product that is bulk-packed (i.e., bulk-cooked meat patties, bulk burritos) will be collected in a slack-filled container. That is, if the product is normally packed in a poly-lined box, the liner will be only partially filled, approximately 1 lb of product, and the liner removed and sealed. This is the product that will be sent to the lab. When conducting this type of sampling, aseptic collection technique (washing hands, use of sanitized gloves) must be followed.

A company may have customers who require finished product pathogen testing, may conduct the testing as part of their environmental program, or may have samples collected by USDA. When finished product testing is conducted for any pathogen, all products from the line, from cleanup to cleanup, become the sampled lot and will be placed on Hold until acceptable results are received [18]. The plant may opt to stop production after the sample is collected to reduce the amount of product retained or may chose to clean and sanitize the line, to provide an intervention step, before beginning production again. If the plant chooses to conduct a full cleanup, make sure that product or production lines in proximity are protected from over-spray. If the product sample is collected by the regulatory agency, the HACCP preshipment review must be conducted and signed before the inspector can send out the sample. It is also prudent for the company to make sure that all the product is under their control when the sample is collected and that none of the product has already begun to leave their control. In this instance, the inspector should not take a sample but should reschedule collection to a time when product is still under company control. If, however, the company is collecting the sample, it is recommended that they do not conduct and sign the preshipment review. Signing this document indicates that all CCPs have been met and that the product is acceptable for release to commerce. However, until the product has cleared pathogen testing, it may not be suitable for release to commerce, so the prudent approach is to not sign the preshipment review.

Through the efforts of industry and the regulatory agencies, random FSIS samples for the period of January 1 through September 30 showed a 25% decline in *Lm* positives [21]. Companies that take food safety very seriously can be proud of that. In some companies, food safety is just lip service, but companies that walk the walk, beginning with the president and through our operations personnel, have an advantage over the companies that merely test to meet a regulatory requirement. Constant efforts to identify and battle pathogen growth give the company confidence that *Listeria* will not be a problem. While vigilance is what makes the system run smoothly, industry partnerships can help a company put the most effective program together. Along with guidance and advice from trade association members, consultants and FSIS personnel can help interpret the regulations. For those companies with limited internal resources, partnerships can provide peace of mind. The value of verification and validation of sanitation cannot be stressed enough as a valuable part of the total plant food safety system.

References

1. Anon, Innovative Tool Helps Meat Processors Reduce Sanitizer Use By 35%, *Food Safety Magazine*, The Target Group, Glendale, 2004, p. 44.
2. Slade, Peter J., Verification of Effective Sanitation Control Strategies, *Food Safety Magazine*, The Target Group, Glendale, 2002, pp. 24–43.
3. Russell, Scott M., A Mini Guide to Rapid Methods for Monitoring Sanitation, *Food Safety Magazine*, The Target Group, Glendale, 2002, pp. 24–25.
4. Vasavada, Purnenda C., Sanitation Audits: The Proof Is in the Pudding, *Food Safety Magazine*, The Target Group, Glendale, 2001/2002, pp. 22–45.
5. Frey, Teresa, Checklist for Verifying Compliance with FSIS Control of *Listeria monocytogenes* in Ready To Eat Product, National Meat Association Resource, Oakland, 2003.

6. Campbell, Brian, Is Your Plant as Clean as It Looks?, *Food Safety Magazine*, The Target Group, Glendale, 2005, pp. 61–62.
7. Bricher, Julie, Top 10 Ingredients of a Total Food Protection Program, *Food Safety Magazine*, The Target Group, Glendale, 2003, p. 34.
8. Powitz, Robert, Successful Sampling Part 1: Essential Approaches, *Food Safety Magazine*, The Target Group, Glendale, 2004, pp. 62–64.
9. Stier, Richard, Do Your Homework, *Food Engineering*, 2005, p. 24.
10. U.S. Department of Agriculture, *Compliance Guidelines*, Washington, D.C.
11. Anon, Uncover Hidden Costs with System Based HACCP Monitoring, *Food Safety Magazine*, The Target Group, Glendale, 2004, p. 50.
12. Anon, ATP Real-Time Advantage in Sanitation Monitoring, *Food Safety Magazine*, The Target Group, Glendale, 2004, p. 40.
13. Henning, William R. and Catherine Cutter, *Controlling* Listeria monocytogenes *in Small and Very Small Meat and Poultry Plants*, Washington, D.C., pp.
14. U.S. Department of Agriculture, *Listeria Guidelines for Industry*, Washington, D.C., 1999.
15. U.S. Department of Agriculture, FSIS Notice 49-04, *FSIS Form 10240.1 Production Information on Post-Lethality Exposed RTE Product*, Washington, D.C., 2004.
16. Blackwell, John, Guidelines for Environmental Sampling/Testing for Plants Producing RTE Products, excerpted from Guidelines for Developing Good Manufacturing Practices (GMP's), Standard Operating Procedures (SOP's) and Environmental Sampling/Testing Recommendations (ESTR's) Ready To Eat (RTE) Products, North American Meat Processors; Central States Meat Association; South Eastern Meat Association; Southwest Meat Association; Food Marketing Institute; National Meat Association; American Association of Meat Processors, 1999.
17. Anon, Guidelines to Prevent Post-Processing Contamination from *Listeria monocytogenes*, National Food Processors Association, April 1999,
18. Anon, *Listeria Briefing, Requirements of FSIS Directive 10240.3*, Sponsored by the National Meat Association, presented by the HACCP Consulting Group, 2003.
19. Katsuyama, Allen M., *Principles of Food Processing Sanitation*, Food Processors Institute, 1993, pp. 91–97.
20. Anand, Sanjier K. and Mansel W. Griffiths, Advances in Biotechniques Used in the Quality Assurance of Food, *Food Safety Magazine*, The Target Group, Glendale, 2004, p. 22.
21. Savage, Robert A., Next Level Strategies in Meat and Poultry, *Food Safety Magazine*, The Target Group, Glendale, 2004, pp. 56–58.
22. U.S. Department of Agriculture, Listeria monocytogenes *Workshop*, Washington, D.C., 2003.
23. Tompkin, R. B., Control of *Listeria monocytogenes in* the food-processing environment, *Journal of Food Protection*, Vol. 65, No. 4, 2002, p. 720.
24. U.S. Department of Agriculture, FSIS Notice 49-04, *FSIS Form 10240.1 Production Information on Post-Lethality Exposed RTE Product*, Washington, D.C., 2004.
25. Byron, Jim, Rethinking Listeria: Things Change, *Food Safety Magazine*, February/March 2012, pp. 12–15, 64.
26. Gombas, Dr. Dave, Environmental Monitoring for Listeria: Getting Started, *Food Safety Magazine*, April–May 2012, pp. 26–27.

27. U.S. Department of Agriculture, Directive 10245.5, *Verification Procedures for Enforcement, Investigation and Analysis Officers (EIAO's) for the* Listeria monocytogenes *Regulation and Routine Risk-Based* Listeria monocytogenes (RLm) *Sampling Program*, Washington, D.C., 2009.

28. U.S. Department of Agriculture, *Notice 03-12, FSIS Actions in Establishments that Temporarily Alter Routine Practices during Routine Risk-Based* Listeria monocytogenes (RLm) *Sampling or Intensified Verification Testing* (IVT), Washington, D.C., 2012.

Chapter 9

Employee Good Manufacturing Practices

... there was simply no such thing as keeping decent, the most careful man gave it up in the end and wallowed in uncleanness. There was not even a place where a man could wash his hands, and the men ate as much raw blood as food at dinnertime.

The Jungle by Upton Sinclair

Rationale for Good Manufacturing Practices (GMPs)

Once the plant has been cleaned and sanitized, it is important to maintain this condition during production, when there are exposed ingredients, in-process materials, and finished products present. Plant personnel are one of the most significant potential sources of microbiological, physical, and chemical food hazards. Because employees frequently move throughout the manufacturing facility, they take with them the potential to spread food hazards from area to area. For example, employees working in a raw meat area can track spoilage or pathogenic bacteria to the cooked product department, or employees wearing uncovered jewelry can be a source of physical contaminants. For this reason, persons working in direct contact with food, food contact surfaces, or food packaging materials must conform to hygienic practices while on duty. These hygienic practices include, but are not limited to, hand cleaning, the wearing of hair covering and clean outer garments, jewelry control, food and tobacco control, disease control, and control of employee traffic.

In order to document the requirements for personnel practices, food manufacturing facilities will establish and implement a Good Manufacturing Practices procedure, commonly referred to as GMP policy. The GMP policy shall be based on the local code of regulations—this would be the Code of Federal Regulations (21 CFR, part 110.10) in the United States. The FDA uses GMPs to control contamination of foods by preventing introduction of microbiological, chemical, and physical contaminants [1]. This regulation sets requirements for food processing operations, including employee training, facility design and construction, maintenance, sanitation, operations, testing, and record keeping [2]. The plant GMP program is written to include guidelines for employee requirements, expectations for visitors, and an effective self-assessment program to verify compliance. Once GMPs have been presented to employees, through initial training, they are expected to follow them to provide a measure of product protection. It is of utmost importance that plant management set the example for implementation of GMPs. They must follow the policy without exception if they expect that all plant personnel, including sanitation and maintenance, will comply with GMPs to prevent product contamination during manufacturing. All visitors to the plant, contractors, suppliers, and regulatory personnel, will be informed of and expected to follow plant GMPs to prevent incidental process or product contamination. In most instances, regulatory personnel will follow GMPs if instructed to do so and are expected to do so by their supervisors. If compliance is an issue, take it to the next regulatory level for enforcement.

The following are some of the more common compliance criteria for food plant GMPs. While all of these may not apply to every food plant operation, and some may not be included due to the specifics of additional customer requirements, they can be reviewed for application in each manufacturing environment.

Basic Good Manufacturing Practices

The GMP information listed below is a compilation from many of the references used in writing this chapter and years of working in the food industry. Many of these are basic and common to all references [1,3–7].

Hands and Hand Washing

According to the Centers for Disease Control and Prevention (CDC), hands are the second-leading cause of foodborne illness and that hand washing is the single most effective means of preventing food contamination. Approximately 20% of foodborne illness outbreaks can be traced to infected employees through the fecal–oral route due to food handlers not washing their hands after using the bathroom or handling raw ingredients [12]. Organisms that cause foodborne illness such as *Salmonella, Hepatitis A, E. coli* O157:H7, and *Staphylococcus aureus* can all be traced to the hands and skin. For this reason, hand washing is one of the most critical GMPs and one of the most important to enforce.

In the food manufacturing environment, it is vital to expect all persons entering food manufacturing and handling areas to wash their hands. To accomplish this, it is important to provide a sufficient number of hand wash and hand sanitizing stations to accommodate the number of people going to the line to prevent production delay. Hand wash stations will be conveniently located throughout the plant, where they are in the path of personnel travel and in work pattern areas, not just where space exists. As an example, locate a sink or sinks near the time clock or where smocks and hairnets are passed out when employees enter the processing plant. It is especially important that the washing facilities are positioned prior to the return from the restroom to the production line [4]. They should have hands-free operation (knee, foot pedals, or automated with electronic eye) so that employees do not have to handle knobs, and they will have suitable drying devices, and be stocked with an adequate quantity of soap. Signage will be present to differentiate hand wash sinks from equipment/utensil wash or meat wash sinks. The hand wash sinks should not be used for meat or utensil wash and, conversely, the meat/utensil wash should not be used for hand washing to prevent possible cross-contamination.

Signs shall be located in the processing areas, which direct employees to wash and sanitize their hands before work, after each break, and when their hands become soiled or contaminated due to handling raw meat, pallets, or materials on floors (see Figure 9.1).

Employees shall wash hands (including fingernails) with soap and hot water, and use hand sanitizer prior to commencing plant functions and as often as may be required to remove soil and contamination. They should be trained to understand the importance of washing their hands after touching their face, blowing their nose, coughing or sneezing, scratching their head, or any other activity that might result in bacterial contamination of their hands or fingers. When training employees to wash their hands, explain that it is much like cleaning the plant: rinse, soap, scrub, and sanitize as illustrated in Figures 9.2 through 9.5. Step 1 is to rinse their hands with warm, clear water. Step 2 is to apply soap and scrub the hands. Step 3 is washing hands, including the palms, the backs of the hands, between the fingers, the nails, and cuticles. All of these areas need to be cleaned for hand washing to be the most effective. Work the lather from the soap into the hands so that it will penetrate into the pores of the fingers and hands. Step 4 involves rinsing the hands again with clear water to remove all traces of soap and suspended soil.

When training employees to wash their hands, it is most effective to demonstrate first, have them wash their hands while you observe, and then provide guidance or correction as needed. Exposure time to the soap is also important, at least 20 seconds, and some plants train their employees to sing a song like "Happy Birthday" while washing hands!

Figure 9.1   Signs such as this will be present at all hand wash sinks as well as in restrooms to remind employees to wash hands before returning to their workstation.

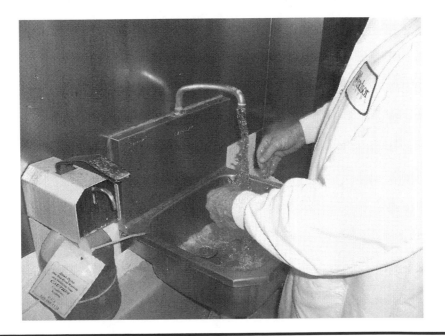

Figure 9.2   Hand washing Step 1: rinse with warm, clear water.

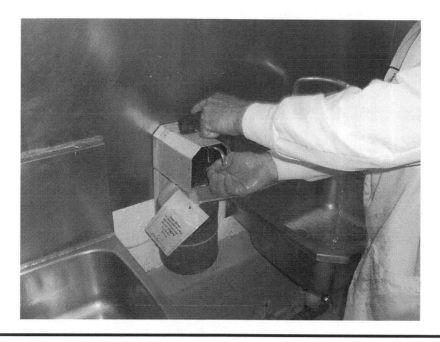

Figure 9.3   Hand washing Step 2: apply soap.

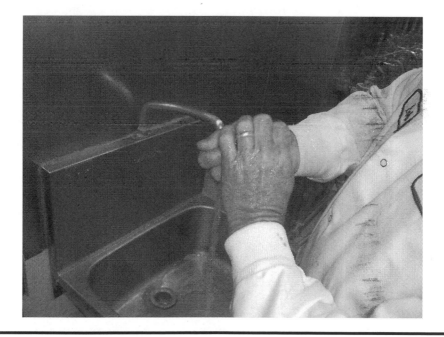

Figure 9.4   Hand washing Step 3: scrub hands with soap, including palms, backs, and fingernails and cuticles.

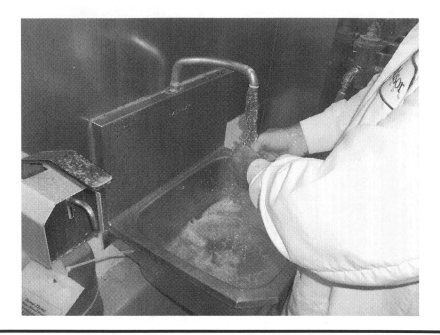

Figure 9.5   Hand washing Step 4: rinse off all soap with warm, clear water.

Hot water provided at wash stations is defined as water that reaches 110°F within 30 seconds. The FDA requires water temperature of 100°F for hand washing, and this is in the Food Code. USDA does not list this as a requirement but as a good guideline for effective cleaning [10]. Use of a bacteriostatic soap, such as a USDA E2 rated soap, is strongly recommended. Once hands are washed, they should be dried with single-use disposable towels or electric air dryers. Continuous roll cloth towels can be a source of contamination because they absorb moisture and roll up against themselves. If they are not advanced by each user, the subsequent user may be drying their hands with a contaminated surface. They are not acceptable for use and must be prohibited from use. Be sure to provide sufficient waste containers for paper towel disposal at sinks so that towels are not thrown on the floor (see Figures 9.6 and 9.7).

Sanitizers for hands after washing range from chlorine to quat, and iodine to alcohol. Care must be taken to ensure that the hand sanitizer provides the necessary level of bacterial destruction without creating skin irritation. Chlorine sanitizers can be harsh on the skin, lead to drying and cracking, which then can lead to sores and infection. Quat (at a level of 150–200 ppm) and iodine (at a level of 25 ppm) seem to result in less irritation but have to be checked regularly during the production shift to ensure that they are sufficiently strong. They may be checked using the appropriate type and level test strip or with titration devices (see Figure 9.8).

Hand sanitizer dips should be replaced if they begin to fall below 75% of their effective level. If used in hand dip tubs, the tubs must be labeled to identify contents

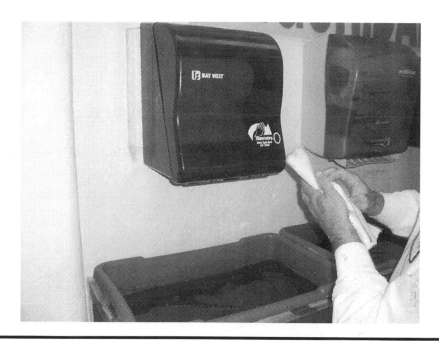

Figure 9.6   After washing hands, dry them with a clean cloth towel or use an automatic air dryer.

Figure 9.7   Provide employees with sufficient trash receptacles to dispose of used paper towels.

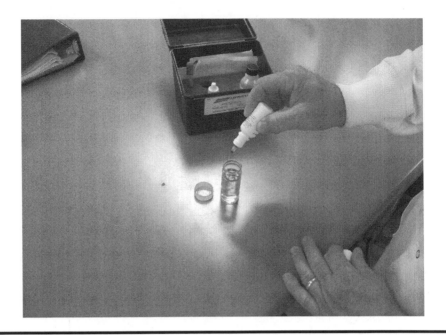

Figure 9.8   A titration kit is one method of checking the strength of sanitizers in the plant. Test strips may also be used, depending on the sanitizer and the concentration.

(i.e., "Quat Hand Sanitizer") and must be cleaned and sanitized daily as with other plant equipment and utensils. When choosing hand sanitizers, consider the organisms to be controlled and plant conditions that may impact employee skin condition. In addition, remember that iodine can result in staining of porous types of containers, which does not reduce the sanitizer efficacy; however, some regulatory individuals may balk at discoloration even if it is resulting from sanitizer (see Figure 9.9).

Alcohol sanitizers, with skin softeners such as lanolin, seem to be preferred for their effectiveness and ease of use. They have been shown to be effective in reducing levels of bacteria and dry quickly, and thus are less irritating to the skin. Because they can be purchased premixed, they do not require monitoring for effective level (although they do need to be monitored and replaced when they run low). They can be wall-mounted with hand operation levers or used with floor stands that can also speed people through the hand wash and sanitizing process, preventing delay in getting to their work areas.

Getting people to wash their hands is very important, and there are several approaches that may be taken to verify that this is being done. The simplest means is to station a production supervisor or QA inspector at the sink location as personnel return to food handling areas from break or lunch and visually observe as they wash. One plant that the writer has worked with uses video cameras located in the plant to verify hand washing is being done. Though some view this as bordering on "Big

Figure 9.9 Signs such as this are good visual reminders to employees to sanitize their hands, even if they are wearing gloves.

Brother," it is simply a tool to ensure compliance. Visual verification allows for an opportunity to take corrective training action if it is observed that employees are not washing their hands, or washing them properly. Other available technology includes the use of swipe cards or radio frequency identification tags to activate sinks and automatically log compliance providing a record for each employee of the number of times that they wash their hands. This log can then be evaluated to identify whether employees are washing hands as required or need further training or discipline.

It is important to verify hand sanitation effectiveness through collection of hand swabs, usually indicator organisms such as coliform or possibly *Staphylococcus aureus*. On a regularly scheduled basis, employee hands (including fingertips and nails) can be swabbed as they return to their workstations to assess how thoroughly they are washing their hands (see Figures 9.10 and 9.11).

This data can be charted and used to determine whether the soap, sanitizer, and hand washing techniques are effective in removing soils, generic bacteria, and potential pathogens. It can also be used to identify personnel that will need additional training on proper hand wash techniques.

Employee Welfare Rooms

Before entering the production plant, many employees start their shift in the plant locker room and connected welfare room. These rooms provide areas where

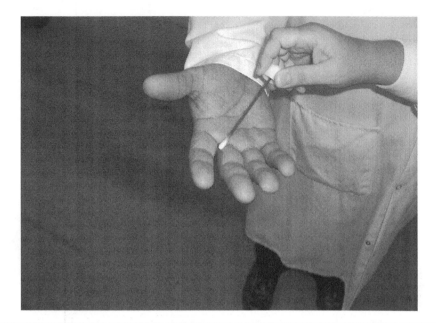

Figure 9.10   Taking hand swabs are a good means of monitoring the consistency and effectiveness of hand washing.

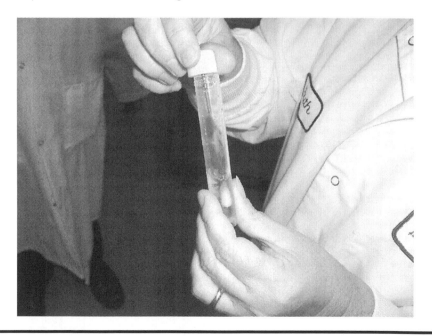

Figure 9.11   After the swab is collected, it is returned to the media tube for incubation.

employees can prepare for their work shift in a clean, comfortable, and sanitary environment. They are usually connected to welfare areas for personal use. It is important for sanitary purposes to maintain welfare areas in a clean condition to prevent potential contamination from these areas from entering the production area and to prevent conditions that might attract pests. Construction of the welfare rooms should be such that they do not open directly onto the production area and the air pressure should be negative in washrooms to prevent potential contamination from entering the production facility.

Welfare rooms will be washed, sanitized, and maintained in a sanitary condition during production when they are most often being used. Strict requirements must be established to maintain these conditions. Before entering welfare areas, employees must be trained to remove outer work clothing such as frocks, aprons, and gloves. Ideally, hairnets will be removed if there is a means of storing them to prevent contamination.

Because many food plant employees come from areas where plumbing is not as efficient as in the United States, they must be trained periodically that toilet paper is to be flushed down the toilet after use. It is not acceptable, and highly undesirable, for used toilet paper to be placed on the floor, or in "waste" containers. Hand wash signs must be posted in the welfare areas in all applicable languages or as pictorials instructing all employees to thoroughly wash and sanitize hands after visiting the restroom and before returning to work.

Locker Storage and Sanitation

Locker rooms provide plant employees with a secure area to store coats, shoes, and personal items restricted by GMPs such as jewelry. It will also provide them with an area to prepare for their shift, so it is important to provide them with a changing room that is both comfortable and well equipped. In order to keep production and ingredient, packaging, and food storage areas free of personal items, employees should be provided with personal storage areas or lockers. Lockers should not, however, be used for storage of food, drink, or production equipment as this may lead to pest infestation and insanitary conditions. Cleanliness of the lockers and locker area is an integral part of the plant sanitation and housekeeping program to promote a food safe environment. If provided, lockers require regular cleaning and Plant Sanitation will ensure that locker cleaning is part of the Master Sanitation. Regular inspection will be conducted to ensure that there is no food stored there, that they are maintained in a clean condition, and that no production equipment is stored there. During these inspections, report any findings of insect or rodent activity to the licensed Pest Control Applicator for corrective actions. In addition to regular cleaning and inspection, it may be necessary to conduct periodic fogging to prevent the growth of pest insects or eliminate those that may be present. Fog lockers with a nonresidual insecticide, never with a residual product, then clean and sanitize them afterward.

Quality Assurance, along with a cross-functional management team and sanitation, will arrange for frequent, announced locker inspections. As with all other GMPs, plant management must ensure that the locker storage and inspection policy is effectively communicated to all plant employees. Announced locker inspections will be conducted on a (minimum) monthly basis. Plant employees should be notified in advance to remove all personal items and locks the night before the inspection. Locks not removed will be cut off, and personal items removed and stored in a plastic bag identified with the employee name. Document all findings, including storage nonconformance, damage, or evidence of insects or rodents. Findings should be reviewed with the plant manager and HR manager for appropriate corrective action, including retraining of plant employees.

Gloves

Use of gloves for food handlers is a subject of debate; however, it is highly recommended that employees who handle food products should wear gloves as an extra barrier of protection between the clean, sanitized hands and food product. As indicated in the section on hand washing, almost 20% of foodborne illness outbreaks can be traced to contamination from the food handler's hands. So the thought process often is cover the hands with gloves, and the problem is eliminated. It may be easy to get lulled into the false sense that gloved hands are totally safe. However, if the gloves become contaminated, through touching raw material, the floor, or the skin, or if the gloves tear, the protection offered can be diminished. Employees sometimes store personal gear such as gloves in places where it is convenient for them such as on top of electrical panels or between walls and piping, but these areas may not be sanitary. Thus, rules of use must be in place if gloves are to be worn by food handlers.

Glove users must wash and sanitize their hands before applying gloves, and gloves are not to be used as an excuse for not washing hands. If used, the gloves must be washed and sanitized as with hands (see Figure 9.12).

Disposable latex gloves will be changed when they are torn or damaged, after absence from the workstation or when potential contaminants are handled. If single-use gloves are used, they must be disposed of when soiled and not reused [9]. Nondisposable rubber gloves must be washed and sanitized after breaks, after handling potential contaminants, or as frequently as needed. Cloth gloves may be worn but must be covered by a latex/rubber glove. Whatever type of glove is chosen, be sure to train employees not to store them in locations that are likely to result in contamination.

Select gloves that are durable enough for the work tasks to be done so that they do not rip or tear. In an effort to prevent potential physical contamination, consider using colored gloves so that the material will be more easily detected in food if they tear or get into the food process flow. Additionally, employees or visitors with nail polish or fake nails on the fingers will be required to wear latex/rubber gloves.

Figure 9.12 Employee wearing gloves applies sanitizer before going to their workstation.

Though this is mostly to prevent foreign material (false nails) from entering the product stream, false nails can also be a source of bacterial contamination as nails and cuticles are difficult to clean.

Hair and Beard Covers

All persons, both associates and visitors, in or near processing, manufacturing, packaging, and warehousing areas, are required to wear clean hairnets to cover head hair. Even those employees who are balding or choose to shave their heads should be required to wear a hair net to ensure consistent application of this GMP in the plant. The best type of hair cover is closed white covers, not mesh or dark covers. The closed white covers ensure that hair will not get out between the strands and make it easier to identify that the person is wearing the net. A clean hair net must be worn in such a manner so as to contain all hair. With some individuals, two nets may be required to cover enclose all of their hair. If the plant requires the use of earplugs for noise abatement, it is best to cover the entire ear and plugs with the hair net. Hair nets and beard covers are not to be worn outside the manufacturing plant.

In addition, those persons with facial hair (i.e., beard, goatee, lipstache) will wear beard nets. Hair and facial hair (i.e., beard) (see Figure 9.13) must be clean and neatly trimmed for safety and sanitary purposes. In an effort to provide a

Figure 9.13 A plant employee shown wearing designated plant attire: hair net, beard guard, earplugs (for safety), and a colored smock for raw processing (brown).

consistent application of the policy, limitations for facial hair length and coverage should be provided. A clean facial hair (i.e., beard) net must be worn in such a manner so as to contain all facial (i.e., beard) hair with the exception of eyebrows, eyelashes (do not allow false eyelashes as they may fall off and contaminate product), and small mustaches. A small mustache constitutes the upper lip area only and does not require a cover. Facial hair in excess of the upper lip area shall be covered with a beard net. Mustaches may be trimmed neatly and above the corners of the mouth, otherwise wear a beard guard. Sideburns that extend beyond the lobe of the ear should be covered with a beard guard.

Clothing and Footwear

Although it is desirable for all plant employees to wear clean clothing into the plant for their work shift, there is no guarantee that street clothing is sufficiently clean to prevent product contamination. Plants may provide clean uniforms or smocks for employees working with or around ingredients or finished products and require visitors or contractors to wear smocks in processing, manufacturing, packaging, and warehousing. Smocks, where required, must be clean and hygienic. Companies may want to consider if they want to control cleaning of smocks or contract with an outside service for the smocks to be cleaned, sanitized, and delivered to the plant.

Smocks should have snaps for closure and not buttons that can come loose and create a physical hazard in product. When employee's smocks become excessively soiled with food or other substances, they will replace them with a clean smock. Smocks that become torn or shredded should be removed from use to prevent string or thread from entering product. Employees involved in processing, manufacturing, packaging, and warehousing will not store any items in shirt pockets or any other pocket above the waist. Smocks will not have any pockets above the waist so that items that are commonly stored there, such as pens or thermometers, do not fall out into product streams. Any smocks with a pocket above the waist will have such pockets sewn shut or removed. As with hair and beard covers, smocks will be removed when leaving the building and will not be worn outside the plant. The wearing of shorts or skirts should be discouraged for all plant personnel regardless of whether a smock is worn. Skin has a microbial flora that may carry both spoilage and pathogenic bacteria, so it is important to keep as much of the skin area covered on plant personnel as possible. The use of color-coded smocks to differentiate job areas is a good idea to provide visual differentiation of employees in the plant (see Figure 9.14). Darker colors should be discouraged as they do not always show when they are soiled.

In many fully cooked RTE meat and poultry plants, the use of aprons and plastic sleeves are required as an extra means of protection. Gloves, aprons, and sleeves must be maintained in a clean and sanitary condition during production

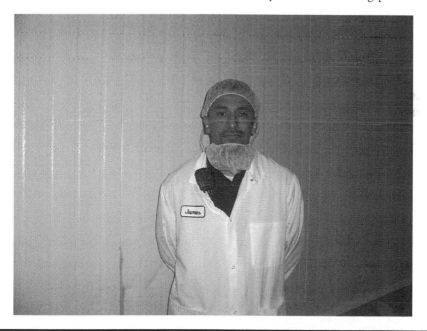

Figure 9.14　Plant employees wear a different color smock for work in Ready To Eat production areas (white).

and should be changed frequently or as they become soiled. If the aprons are the disposable type, they will only be used for one production shift and will then be disposed of. If they are the kind that can be reused, they will be washed, sanitized, and dried at the end of the shift before being hung in the employee locker. Plastic sleeves, if worn, are also to be maintained in a clean and sanitary condition during the shift.

Appropriate foot covering will be worn in processing, manufacturing, packaging, and warehousing areas. Foot covering will protect the plant from contamination as well as protect the associate from injury. Open-toed shoes, high heels, or sandals are prohibited for employee safety as well as for sanitary reasons. Employees will sanitize their boots prior to proceeding to their workstations.

In most cooked RTE meat and poultry plants, there is a captive shoe policy. This policy requires that plant employees wear shoes that meet certain criteria for food and for personal safety. The requirements include defining footwear as no canvas or leather shoes, sneakers, open-toed shoes or sandals, no porous material. It is best to have footwear that is water resistant for comfort, especially where foot dips or floor foamers are used. Rubber boots with good traction and cleanable bottoms are also preferred for personnel and food safety. Captive boots may be stored in lockers if they are provided, but they must be cleaned and sanitized prior to storage.

An added measure of protection in the plant and part of overall sanitary plant practices includes the use of floor sanitizer foamers or sanitizer footbaths (dips). The objective of foamers or dips is to provide a means of sanitizing shoes or boots of employees and visitors before they enter exposed product areas. Foamers are preferable over footbaths because footbaths can build up food and scrap material over the course of a day and have often been described as "bacterial soup." Once organic material builds up in the dips, their effectiveness is reduced. If they are not changed, cleaned, and recharged with fresh sanitizer on a regular basis, they can create more of a bacterial hazard than they prevent. If used, they should be deep enough that they will cover the entire sole of the shoe and halfway up the foot. The sanitizer level should be between 800 and 1000 ppm, and quat is preferred in operations where *Listeria* is a concern. In operations where *Salmonella* is a greater concern, chlorine would be the recommended sanitizer. One positive comment about boot dips is that they often have rubber tips at the bottom of the dip that will help remove material that collects in the tread of shoes and boots.

Floor foamers are most effective if they are placed strategically in the plant and operated correctly. The best locations for floor foamers are where employees or equipment enter production areas and where there is crossover traffic between raw and cooked areas. The floor foamers should be set to deliver 800–1000 ppm sanitizer and aerated to form thick, wide foam. The foam should be wide enough that a person cannot step over the stream and so that, as wheeled vehicles pass through the foam, they will complete at least one revolution. The foamers should be timed to go off frequently enough that there will be a consistent quantity of foam during the production hours, especially during heavy foot and wheeled traffic times.

Figure 9.15   Effective implementation of a floor foamer in a plant. The foam crosses the entire path of employee travel after washing and sanitizing hands.

It will be prudent to evaluate floor condition when deciding between using floor foamers and foot dips. Floors should provide traction so that the wet conditions that will be created from the sanitizer do not pose a safety hazard to employees (see Figure 9.15).

Jewelry

Jewelry control is done more for prevention of foreign material contamination than for sanitary purposes, but no GMP procedure will be complete without at least addressing this. However, bacteria can hide under watches and bracelets making them a potential source of contamination [11]. Jewelry for purposes of GMPs includes wristwatches, earrings, necklaces or chains, and rings, and should be prohibited in any processing, manufacturing, packaging, and warehousing area by any employee or visitor. Some plants allow rings without stones; however, some bands have texture that will be difficult to clean and should be discouraged. A solid wedding band no wider than 7/16 inches and with no stones may be worn. Glove protection may be worn if a wedding band with a stone cannot be removed. One exception to this rule may be Medical Alert bracelets or chains; however, every means of protection should be taken to prevent them from getting into product. Other personal decorative items that should be considered for control are false eyelashes, false nails, or exposed nail polish that can become physical contaminants in food products.

A recent phenomenon, especially among younger employees, is tongue piercing, among various other locations such as lips, noses, and eyebrows. Obviously, the exposed pierced parts of the body (lips, nose) should not be allowed in the plant without first removing the piercing. However, what should the plant position be with regard to a pierced tongue? As an industry best practice, employees with pierced tongues should not be allowed to work in the plant. It is best not to allow any piercing that is not covered by clothing and the tongue cannot be covered by clothing. Granted, it can be covered by a beard guard; however, if you would not let someone wear pierced earrings even with a hair net, it stands to reason that you would not allow someone to wear their tongue piercing in the plant even with a beard guard. The key is to maintain a consistent application of the intent of the policy.

Disease Control

In an effort to prevent the spread of illness-causing bacteria into product, no person affected with any disease in a communicable form, or while a carrier of such disease, shall work in the manufacturing plant in any capacity which would bring that person into contact with the processing, handling, storing, or transporting of food products. Any person who has boils, sores, an open lesion, wound, or any other abnormal physical condition that can be related to microbiological contamination shall be excluded from any operation that may result in product contamination. Employees who display symptoms of illness such as fever, sore throat, jaundice, or flu should not be allowed to work with ingredients, food, or packaging until the illness is resolved.

All employees whose hands come in contact with product and product packaging, who have open wounds or sores, will wear adhesive bandages or tape on fingers, hands, or forearms. All bandages must then be covered with a nonporous covering such as latex or plastic glove or a plastic sleeve. It is also recommended that the bandages are a bright color or a color different from that of the food being produced so that it can be easily spotted if they accidentally make their way into product. Some bandages are made with thin metal in them so that they can be picked up by plant metal detectors.

Food and Tobacco

Drinking, eating, gum chewing, spitting, and the use of tobacco in any form (i.e., chewing tobacco) by any person in production, warehouse, storage, and maintenance areas is strictly prohibited during production and nonproduction days. These activities should be restricted to areas specifically identified by plant management for smoking, eating, and drinking. In addition, eating or drinking in the plant on nonproduction days by maintenance or sanitation personnel should also be prohibited. This is to prevent food spills or leaving empty food or beverage containers in production or storage areas that can attract pests and create insanitary conditions.

Raw and Cooked Separation

As a means of preventing contamination of cooked products, employees working in raw product areas will avoid traveling through cooked product areas. Additionally, cooked product employees must avoid traveling through raw product areas if they are to return to cooked product areas. The same will apply to plant equipment (i.e., conveyors, gondolas, vats, pallet jacks, forklift, and metal detectors). Equipment used for raw product should not be used for cooked RTE product, even if that equipment is cleaned in between use. The chances of cross-contamination from raw to cooked are too great if the equipment has cracks, hollow material, hard-to-clean areas, or has the potential for biofilm buildup.

Individuals must not be permitted to move freely from one type of process to another without a garment change where the possibility of cross-contamination exists (e.g., from a raw to a cooked process area), and outer garments must be clean at the start of each shift. One means of preventing personnel from moving from raw to cooked areas, or vice versa, is to have an internal color-coded smock policy for raw and cooked areas. This provides a very good visual indicator that people are where they should be and are not where they should not be. In addition, plant equipment can be labeled designating usage in raw or cooked areas and will not be interchanged between the areas.

Visitors and Contractors

Visitors and contractors are also required to comply with company GMPs if they are to enter the production plant. They should be restricted from entering production areas during processing unless they have been informed of and conform to plant GMPs. Visitors include regulatory representatives; in fact, FSIS employees are required to follow the establishment protocol for GMPs provided they have been informed of the expectations. Do not assume that visitors are fully aware of the requirements or know what practices will protect product. Provide them with training if they are not certain how to wash their hands or wear protective gear.

One means of communicating GMPs to visitors is to have a GMP sign-off sheet that delineates your expectations. By signing the sheet, they are saying that they agree to follow your requirements regarding smocks, hairnets, jewelry, etc. This can be done in conjunction with signing a Non-Disclosure or Confidentiality

TIP: Some auditors coming into your plant will not ask about your GMP requirements before entering the manufacturing area. If you do not tell them what you expect, they will deduct points from the audit for not informing them. Always make sure to tell auditors what they must do. Treat them as you would any other visitor that might not know your GMPs.

document as demonstrated in Table 9.1. Provide visitors or contractors with hairnets, smocks, and captive boots as your plant requires, and show them how they are to be worn. Instruct them to remove all jewelry and show them where and how to wash and sanitize their hands. When in the plant, show them how to use foot dips or sprays if these are present. Unless they are working on a specific piece of equipment, they should not handle product or equipment used for product contact. They should be escorted during their visit for food safety, personal safety, and security reasons.

Maintenance GMPs

Maintenance of facility and equipment is a daily necessity to keep a plant operating efficiently and safely. Maintenance personnel work on equipment at all times in the plant and in some instances assist the sanitation department with tearing down equipment for cleaning or reassembling equipment for production start-up. They may also work on equipment during production. Because of the nature of maintenance work, there is the potential for them to accidentally contaminate product contact surfaces and thus product. In order to avoid this type of contamination, it is for maintenance employees to follow all GMPs required by product handling personnel.

There is also a need for maintenance to have their own set of GMPs related to their job and their equipment to minimize potential for product contamination.

- Personnel: Obviously, maintenance personnel should wear hairnets and smocks when working in production areas. They should be provided with separate, unique-colored smocks for maintenance, or be required that they wear the color or designation of the department they are working in. There may be a need for maintenance employees to work on low portions of equipment or even underneath equipment; however, they should not be lying directly on the floor. To accomplish this, it is important to place a barrier, such as a plastic tarp or layer of cardboard, between the maintenance employee and the floor to avoid picking up contaminants from the floor and transferring them to equipment. If possible, designate specific personnel to raw and others to fully cooked RTE areas to avoid cross-contamination. Where this is not possible, maintenance must follow very strict decontamination procedures when leaving a raw area and entering a cooked area to work, such as hand washing, changing smocks, and cleaning and sanitizing shoes.
- Tools: Where possible, these should be of a sanitary design as identified in the AMI recommendations. They should never have wooden parts as wood is porous, not readily cleanable, and may have bacterial harborage crevices. There should not be cracks in plastic grips that can harbor bacteria. There should be a sanitary location for storing tools such as a tool locker or tool box. These should be maintained in a sanitary condition and cleaned and sanitized on an established frequency, which means that they must be made

Table 9.1 Cleo's Foods Requirements for Visitors and Contractors

All visitors and contractors to Cleo's Foods plant are to comply with the following policies. Failure to comply with these policies will be considered noncompliance and grounds for removal from the plant or termination of our agreement. Failure to follow Cleo's Foods policies will result in removal from the property. Visitors and contractors will be held liable for any damage to Cleo's Foods property.

FOOD SAFETY

1. **Smocks:** Approved smocks, provided by Cleo's Foods, will be worn in the production and warehouse areas of the plant. Smocks must be removed before going outside, and entering restrooms.

2. **Hairnets:** Approved hairnets, provided by Cleo's Foods, will be worn in the production and warehouse areas of the plant. All hair must be covered. Hairnets must be removed before going outside or entering restrooms.

3. **Beard nets:** As needed, approved beard nets, provided by Cleo's Foods, will be worn in the production and warehouse areas of the plant. All facial hair must be covered. Beard nets must be removed before going outside and entering the restrooms.

4. **Food and drinks:** Food and drinks are not to be eaten, except in approved areas, (the lunchroom). All refuse must be properly disposed of. Gum or candy is not permitted in the production area.

5. **Jewelry:** No jewelry is to be worn in the plant at anytime. This includes watches, rings, earrings or exposed piercings, bracelets, and necklaces. Medic alert jewelry is allowed.

6. **Product contamination prevention:** Contractors are not allowed to touch any product at anytime. Contractors are not allowed to touch any of the production equipment when it is being used for production unless otherwise instructed by the Cleo's Foods representative. No glass is allowed in production areas. This includes glass containers, meters, tools, or utensils that glass are a part of.

7. **Smoking and tobacco:** Smoking or the use of any tobacco product (chew, snuff) is not allowed anywhere in the plant. Smoking is only allowed in designated areas. Do not throw cigarette butts on the ground. The butts must be properly disposed of.

8. **Spitting:** Spitting is prohibited in all areas.

9. **Outside doors:** All outside doors must remain closed at all times and will not be propped opened by contractors.

Continued

Table 9.1 (*Continued*)   Cleo's Foods Requirements for Visitors and Contractors

All visitors and contractors to Cleo's Foods plant are to comply with the following policies. Failure to comply with these policies will be considered noncompliance and grounds for removal from the plant or termination of our agreement. Failure to follow Cleo's Foods policies, they will result in removal from the property. Visitors and contractors will be held liable for any damage to Cleo's Foods property.
GENERAL SAFETY
1. **Lock Out Tag Out:** When working on equipment, it should be locked out and a tag placed on the equipment, indicating who locked it out and why. Make sure that before you leave remove the lock if the job is complete or let a Cleo's Foods representative know the status and when work on the equipment will be finished.   2. **Hearing protection:** Hearing protection must be worn anytime you are on the production floor.   3. **Forklifts:** Contractors are not permitted to use forklifts without the proper training from Cleo's Foods, or written permission from a manager.
SANITARY
1. **Hand washing:** All contractors must wash and sanitize their hands after using the restroom and before entering any production area.   2. **Tools:** All tools used on production equipment must be clean and sanitary before use.
SECURITY
1. All visitors and contractors will check in with the Cleo's Foods representative that requested the work. When they are done for the day, they must check out with that representative before leaving the facility.   2. Visitors and contractors agree that there are formulas, equipment, or processes that are proprietary to Cleo's Foods that will not be disclosed outside the company or to competitors. Picture taking or recording will not be done without prior approval of Cleo's Foods management.
This indicates that the visitor or contractor has read and agrees to follow the policies as outlined on the requirements for Visitors and Contractors for Cleo's Foods.   **Visitor/Contractor Information**   **Company Name:** _____   Company Representative: _____ Date: _____
This form will be kept in the main office for verification and review.

of materials that can withstand water cleaning and sanitizer contact. Some maintenance personnel are concerned that repeated exposure to sanitizer may deteriorate tools. Alcohol wipes can be used as an effective sanitizer that dries quickly and limits corrosion or deterioration. Leather tool pouches should not be used as these are not easy to clean and in microbiological tests have tested positive for environmental organisms of concern. Tools must not be stored in lockers or taken from the plant for personal use. Segregate tools and utensils between raw and cooked so that tools used for raw equipment are not used on fully cooked or RTE product equipment.

- Equipment: Tools and utensils used in the maintenance process require proper storage. Tools coming in contact with product handling or contact equipment should never be placed on the floor as contaminants from the floor may be transferred to the tool and then onto equipment that is being worked on. Parts or equipment pieces should never be placed on the floor when they are removed from larger pieces of equipment for maintenance. Carts should be used for smaller parts or equipment racks for larger pieces of equipment.
- Cleaning and sanitizing: Food contact equipment or surfaces contaminated by maintenance activity before or during operations will be properly cleaned and sanitized before contact with food product or packaging. Some plant operations use a tag system to identify pieces of equipment that have been handled by maintenance. Once the equipment has been cleaned and sanitized, it is inspected by QA before the tag is removed, releasing the equipment to production.

Regardless of the overall procedures required by the plant, make the maintenance department a participant in maintaining a food safe environment. They can be a valuable asset in the prevention of food product contamination.

Training and Implementation

Once the GMP program has been written, it is vital to communicate the program to all plant employees, both current and new employees. Training is a complex process that requires multiple approaches to be effective. It is not enough to simply hand people a copy of the GMPs and expect them to understand and follow, nor is it sufficient to just show a video. Audiovisual aids are an effective tool but, not by themselves; they are best used in conjunction with an explanation as to why GMPs are to be followed and a demonstration as to how they are to be done. This may require a demonstration to show the proper way to put on a smock or a hairnet.

One of the most effective means of training demonstrations is through the use of hand wash kits with fluorescing oil or powder and UV light to demonstrate hand washing effectiveness. The process involves putting the oil or powder on the employee's hands, then having them wash their hands. Once their hands are washed, use the UV light to evaluate the hands. If they were washed correctly and all of the oil or powder is removed, they will not show up under the UV light. However, if they are not washed correctly, the UV light will show where the fluorescing material is still present and further training will be required to ensure that hand washing is effective at removing all of the material.

Maintain records of initial and ongoing associate training with signatures from each employee. Following initial GMP training, it is important to follow up with training approximately every six months. Again, this training should be documented with a sign-off sheet for all participants. Some plants hold daily line meetings just prior to the start of production to review the schedule or productivity requirements, others hold monthly department meetings. These are good time to also review a GMP, especially if there is one found to have frequent deviations. It is also a good idea to post visual aids in employee areas as reminders of Good Manufacturing Practice responsibilities.

Monitoring and Enforcement

During the initial training of personnel, it is important for them to understand that compliance with GMP requirements is essential and that they are ultimately responsible for conformance. It should also be explained that regular monitoring will be conducted to verify compliance. Production leads or supervisors are responsible for monitoring employee conformance with GMPs on a daily basis using a form similar to the example in Table 9.2. Monitoring will provide them with an opportunity to conduct additional training or counseling to ensure that the expectations are clearly understood and to correct behavior that does not conform to expectations. They will also verify that employees are aware that nonconformance to GMPs may result in disciplinary action.

GMP self-assessments should be conducted on a monthly basis to document employee compliance to the policy. While normally conducted by the QA department, a cross-functional team is recommended composed of the following personnel: Plant Manager, QA, Maintenance, and a production employee. The team will tour the plant and evaluate employee conformance with GMP requirements, taking corrective action if deviations are found. All findings or corrective actions will be documented to provide a guide for follow-up training with groups or individuals.

Operational Sanitation

Each USDA-inspected meat and poultry plant is required to make a daily operational sanitation inspection to fulfill requirements of SSOPs. In non-USDA plants, this may be referred to as Housekeeping inspection. Housekeeping and sanitation are integral to the operation and closely linked. These are additional practices that will aid in the prevention of product contamination while maintaining safe and sanitary plant conditions. In general, additional housekeeping practices include some of the following steps to provide for sanitary food plant operations:

Table 9.2   GMP Monitoring Checklist—Example

Cleo's Foods GMP Monitoring Checklist				*Date*	
The checklist is to be completed by Quality Assurance a minimum of twice per shift. Place a "Y" for "Yes" or an "N" for "No" in the block after the GMP comment. A "No" response for any item will require a comment in the section below. Two "No" responses will require further action by QA up to and including production stoppage until the deficiency is corrected. Have the supervisor/line lead sign after each check.					
		TIME #1	TIME #2	TIME #3	TIME #4
		#1 *Y/N*	#2 *Y/N*	#3 *Y/N*	#4 *Y/N*
1	Water temperature is 110°F within 30 s.				
2	Adequate soap, paper towels, and trashcans are present at sinks.				
3	Employees wash their hands when returning to work.				
4	Hand sanitizer strength is 150–200 ppm (50 ppm for iodine).				
5	Hairnets and beard nets are properly worn.				
6	Jewelry is removed and not worn in the plant.				
7	Gloves are worn by product handlers and are clean.				
8	Aprons are worn and are clean.				
9	Smocks are worn and are clean.				
10	Appropriate, clean shoes are worn.				

Continued

Table 9.2 (*Continued*)   GMP Monitoring Checklist—Example

Cleo's Foods GMP Monitoring Checklist			Date	
The checklist is to be completed by Quality Assurance a minimum of twice per shift. Place a "Y" for "Yes" or an "N" for "No" in the block after the GMP comment. A "No" response for any item will require a comment in the section below. Two "No" responses will require further action by QA up to and including production stoppage until the deficiency is corrected. Have the supervisor/line lead sign after each check.				
	TIME #1	TIME #2	TIME #3	TIME #4
	#1 Y/N	#2 Y/N	#3 Y/N	#4 Y/N
11 Floor sanitizers are operating and at proper levels.				
12 Food, beverage, and tobacco use is restricted to specified areas.				
13 Raw and cooked employees stay in designated areas.				
14 Employees appear to be free of illness or infections.				
15 Maintenance is following all GMPs.				
16 Maintenance personnel keep tools off the floor.				
17 Locker rooms are free of food or beverages.				
18 Visitors (if present) are appropriately attired.				
19 Supervisor/lead initials.				
COMMENTS:				
AUDITOR: Confidential Commercial Information				

- Ingredient spills will be cleaned up quickly in storage areas to prevent accumulation that creates insanitary conditions and can attract pests. Product spills in the manufacturing areas will be removed to prevent buildup that creates insanitary conditions or personnel safety hazards.
- Authorized sanitation personnel cleaning the plant during production and using water hoses will avoid splashing water from unclean surfaces (such as floors and drains) onto exposed product or product contact surfaces. If possible, reduce the water pressure to prevent overspray from product surfaces or creation of atomized particles from drains.
- Production sanitation personnel responsible for emptying trash, cleaning floors, and picking up scrap will be trained to avoid touching product or product contact surfaces.
- All product packaging, both primary and secondary, and packaging supplies will be used only for packaging. Similarly, food ingredient or in-process material containers must be used only for ingredient or food product storage. They will not be used for chemicals, washing compounds, spare parts, tools, or any other similar nonfood use (i.e., paper, form storage, etc.). This is necessary to prevent physical or chemical contamination of ingredients or product.
- Containers bearing compounds (i.e., cleaning chemicals, sanitizer, white oil, etc.) shall be identified as to contents as shown in Figure 9.16. Use dual language (English/Spanish) as necessary, depending on plant associate language base. Unlabeled containers will be removed and disposed. Again, this is done to prevent chemical contamination of product (see Figure 9.16).
- Walking on or climbing over ingredients or product will be prohibited. This includes packaged goods or exposed ingredients or product. Because of space constraints, some plants have steps or access over product streams. These need to be designed in a way that they will not contribute to product contamination.
- All idle equipment stored within the plant or auxiliary buildings shall be covered or otherwise protected to prevent the access of dust, condensation, or other contamination. If they are to be used for production, they must be fully cleaned and sanitized prior to use. If they have been stored in an off-site location, inspect them as they are brought into the plant for any damage or design that might contribute to unsanitary conditions.

As with any other plant food safety or quality program, the GMP program will be updated periodically to reflect emerging food safety hazards or changes in the plant products or processes. If properly developed and implemented, GMPs can be an extension of the sanitation program by ensuring that product or process contamination does not occur through employee actions after the plant has been cleaned. Getting employee support and cooperation is critical, and monitoring of practices can provide assurances that the program is working as designed.

Figure 9.16 Containers for chemical compounds, in this case sanitizer for storing a thermometer, are clearly identified as to their contents to prevent chemical contamination of product.

References

1. Gould, Wilbur A., *CGMP's/Food Plant Sanitation*, CTI Publications, Baltimore, 1994, chap. 18.
2. Stauffer, John E., *Quality Assurance of Food Ingredients, Processing and Distribution*, Food and Nutrition Press, Trumbull, 1988, p. 119.
3. Frederick, Tammy, The Do's and Don'ts of Food Plant Personal Hygiene Practices, *Food Safety Magazine*, The Target Group, Glendale, 2004.
4. Powitz, Dr. Robert, A Practical Perspective on Hand Washing, *Food Safety Magazine*, The Target Group, Glendale, 2003, pp. 16–17.
5. U.S. Food and Drug Administration, Current Good Manufacturing Practices, 21CFR110, FDA Website.
6. Keller, JJ & Associates, *Employees Guide to Food Safety*, 1998.
7. AIB Consolidated Standards, Food Safety, American Institute of Baking, Manhattan, KS.
8. Imholte, Thomas J. and Tammy Imholte Tauscher, *Engineering for Food Safety and Sanitation, A Guide to Sanitary Design of Food Plants and Food Plant Equipment*, Technical Institute of Food Safety, Medfield, 1999.
9. Michaels, Barry, There's More to It at Hand, *Food Quality Magazine*, 2005, p. 71.
10. Anon, What Is HCC?, *Meat Processing*, 2005, p.22.
11. DeSorbo, Mark, Hands Free for Hygiene, *Food Quality Magazine*, 2005, p. 23.
12. Anon, From Cuticles to Counter Tops: Alcohol Sanitizers Aid Hygiene, *Food Safety Magazine*, The Target Group, Glendale, 2005, p. 30.

Chapter 10

Pest Control and Sanitation

... but with the hot weather there descended upon Packingtown a veritable Egyptian plague of flies; there could be no describing this – the hoses would be black with them.

The Jungle by Upton Sinclair

It is fortunate that conditions such as the one described in Upton Sinclair's classic novel about the meat-packing industry can be easily be avoided given all that has been learned over the years about sanitation and pest control. The food plant environment is attractive to pests as it provides ideal conditions and basic survival needs for pests: food, water, warmth/temperature, security, and absence of natural predators. Insects and rodents are attracted to odors from food plants and lighting used both inside and outside the facility. Pest control is part of the overall sanitation plan; in fact, a high percentage of pest control is sanitation. It is intended to prevent contamination of ingredients and food products and is a requirement for conformance to federal regulations. The Code of Federal Regulations, Current Good Manufacturing Practices (21CFR110.35) makes it very clear that pests are to be excluded from food plants:

> (c) Pest control. No pests shall be allowed in any area of a food plant. Guard or guide dogs may be allowed in some areas of a plant if the presence of the dogs is unlikely to result in contamination of food, food-contact surfaces, or food-packaging materials.

> Effective measures shall be taken to exclude pests from the processing areas and to protect against the contamination of food on the premises by pests. The use of insecticides or rodenticides is permitted only under precautions and restrictions that will protect against the contamination of food, food-contact surfaces, and food-packaging materials.

Pest infestation results in product adulteration, which can lead to product loss, possible recall or regulatory control action, and potential loss of business. In order to control pests, it is important to know how they function in the environment. Some pests are beneficial to the environment; for example, bees pollinate flowers and trees, and flies serve as food to certain animals, recycle manure nutrients into the soil, and help decompose carcasses. However, it is also recognized that many pests carry disease or spread microbes, damage ingredients, and infest products. For example, mosquitoes carry diseases such as malaria; flies and roaches spread microbes from the environment they feed or breed in; beetles can infest seasonings or flour, rendering them unusable; and rodents and birds can transmit disease and filth.

Some of the most common food plant pests are flies and beetles, roaches, rodents, and birds. The approach most recommended for control of pests is Integrated Pest Management (IPM). IPM is a multiple hurdle approach that provides several barriers to thwart pests. It begins with the recognition that pests have three basic needs to survive: food, water, and shelter [8]. IPM uses approaches to control conditions that will allow pests to survive. These approaches include elimination of breeding and harborage, exclusion from access to plant, sanitation, and extermination. One advantage of IPM is that insecticide use is a treatment of last resort, and thus chemical use should be minimized.

As a means of facilitating pest control, learn about the specific pests in your region and about the conditions that pests tolerate. For example, insects cannot thrive in the cold (≤40°F), but rodents can tolerate cold temperatures, even inside freezers. Some insects have specific habitats (bee hives, bird nests), while others have no specific habitat but adapt to the environment encountered. This chapter will identify some of the pests of concern to food plants and discuss control measures focusing on sanitation.

Pests of Food Plant Concern

Insects

It is estimated that there are between 800,000 and 1 million species of insects on earth. As stated earlier, many of them are significant and necessary in the environment. However, there are many pests that are of concern to food plant manufacturing. The most common include flies and flying insects, roaches, and beetles. Flies are seasonal in most parts of the country, and the housefly (*Musca domestica*) is the most common fly in the world due in part to its being a prolific breeder. They can be found almost anywhere, but prefer warm temperatures. Adult flies feed on a variety of food, garbage, and manure; thus, they can carry many disease-causing organisms. They use a sponging mouthpart to vomit on a food source. This begins to dissolve the food, which they then suction up using the mouthpart. Signs of flies, other than visual sightings, are dark specs indicating fecal material. Congregation areas and their concentrations can be identified by placing white paper on walls or beams for a period of time. As they settle on the paper, they will leave trace specs to indicate their presence. Control of flies includes exclusion and elimination of food and ingredient waste. Sanitation is important through the removal of debris or food buildup under or around equipment legs or frames. Fruit and phorid flies can breed in large numbers in the presence of accumulation of moist organic debris, such as in drains. Organic matter film must be removed from drains to eliminate fly larvae. Merely treating the drain with hot water or chlorine is not enough [1]. Breeding spaces in walls and pits must be dried and the source eliminated coupled with Ultra Low Volume (ULV or space) treatment with non-residual to kill adults. Termites, while not a major threat to food, can cause severe damage to building structures. The primary indicators of termites are swarms that only last a few days. Once the swarming has stopped, it is important to find their nests and eliminate them at their source.

In the disaster film *The Day After*, broadcast in 1983, the effects of nuclear war on a rural Kansas town are portrayed. One of the scenes involves the late actor Jason Robards looking at a cockroach and commenting that they were on earth before man and they will likely be on earth after man is gone. Though roaches may not survive a nuclear holocaust, this scene somewhat exemplifies the idea that roaches are prolific and hard to eliminate, and this is one of the reasons why they are a challenge for the food industry. The three most common types of roaches found in the United States are the German roach, the American roach, and the Oriental roach (see Figure 10.1). They differ in both color and size. German roaches, the most common, are about ¾ inches in length and are yellow brown in color with two dark stripes behind their head. American roaches are the largest variety, about 2 inches long and brown in color. Oriental roaches are about 1 inch in length and are dark brown to black [4]. All feed on various food sources such as ingredients and finished product, dead insects or other roaches, and human waste. For this reason they are a significant source of bacteria, including pathogens. They prefer to feed in solitude and darkness, so if they are seen during the day, that is a good indication that they have infested an area. Roaches are difficult to eliminate once established, and chemicals that kill adults may not kill the egg cases. The primary means of control are exclusion through inspection of incoming boxes and pallets, treatment and sealing of cracks and crevices where they hide, and elimination of wet areas.

Beetles are another species of insect that are considered beneficial to the environment. Some feed on insect pests of plants, others pollinate flowering plants, and

Figure 10.1 Cockroaches, from left to right, the German roach, the American roach, and the Oriental roach.

many help with the decomposition of plants, animals, or animal feces. They are one of the earliest groups of higher insects to evolve and adapt to the changing environment. Their hard shell provides protection from predators. There are approximately 110 to 115 families of beetles, with the five largest families containing more that 20,000 species [2]. Unfortunately, they are also a pest of crops and stored products. The listing of stored product pests includes the red flour beetle, granary beetle, sawtooth grain beetle, and the Indian meal moth. Once they have been detected in a food plant, it is likely that they have already infested ingredients or foods. If this is the case, then it is imperative to remove the infested materials and dispose of them. Identification of the beetle will aid in determination of the most effective means of control. It is also vital to inspect other materials stored in proximity to determine whether the infestation has spread. If this is not the case, then it is a matter of preventing reintroduction to the environment or elimination if they are elsewhere in the plant. To eliminate a major source of attraction, clean floors, racks, and pallets on a regular basis to remove spilled ingredients (see Figure 10.2). Since they may also be found in walls, air ducts, and on I-beams, it is important to keep these areas clean and free of food dust or residues [2].

Seal floor cracks to eliminate accumulation of spilled materials. Inspect incoming ingredients to verify that they are not infested, and use them on a first-in first-used (FIFU) basis. If they must be stored for an extended period, then inspect them frequently or place them in a freezer if they might become infested. If chemical application becomes necessary, treat cracks and crevices with a residual, and seal

Figure 10.2   A clean and orderly warehouse, with spills cleaned up quickly, provides an environment that will not attract pests.

them after application. Space treatment or fumigation of storage areas or infested goods in bins or on trailers may have to be conducted.

Rodents

From a regulatory standpoint, the presence of rodents or rodent evidence violates section 402(a) of the Food Drug and Cosmetic Act. Even if ingredients or product are not contaminated, they could be subject to regulatory control action if rodent presence is detected. The FDA has reported that approximately 90% of complaints against food establishments are for rodents and that about 90% of those are for mice. From an economic standpoint, rodents are responsible for costly damage to ingredients, products, and the facility. They are carriers of disease, and their presence usually leaves behind excrement and urine. Rats and mice belong to the Order Rodentia and suborder Myomorpha. There are three primary types of rodents that food plants in the United States may deal with [4]. One of two rats is the Norway rat (*Rattus norvegious*) that includes the Norway, brown, barn, sewer, and grey rats. This rat burrows, lives in holes, and has prominent incisors on the upper and lower jaw for gnawing. They are poor jumpers and climbers and generally regarded as clumsy. The second rat is the roof or black rat (*Rattus rattus*). This rat lives above ground in storage spaces, wall spaces, and is an agile climber. The house mouse (*Mus muscullus*) is a problem for both the food industry and for consumers in their homes. This rodent is mostly nocturnal, an excellent climber, and adapts very easily to varying temperature conditions [13]. They are smaller than rats and usually dark gray on the back and light gray on the abdomen with large protruding ears. Because the roof rat and house mouse are good climbers, they can have access to roofs up walls or rain gutters. On the plant roof around air vents, grease and food residue can accumulate and attract rodents. These areas need to be maintained in a clean condition, and gutters will have to be screened to prevent rodents from using them as an access path. Rodent IPM involves sanitation, rodent proofing, and rodent prevention using a combination of external bait stations, on the perimeter of the property and on the exterior of the facility, and internal traps. Signs of rodent entry to the plant include evidence of gnawing, droppings, and smears from body oils as they travel. They also frequently leave urine, which will glow under fluorescent light so use black light to detect.

Birds

Birds provide a benefit to our lives as pets, have aesthetic value, and eat insects. Some birds, however, are pests and present a threat to crops, including grains, fruits, and vegetables. They also carry diseases such as salmonella and fungi that come from droppings, and they carry parasites such mites and ticks [3]. Birds are attracted to facilities because of food spillage and potential roosting sites. Two of the primary bird pests for food manufacturers are pigeons and sparrows. These birds are not solitary and tend to congregate, so their presence may result in buildup of excreta on or below surfaces where they roost. Their habitats include flat surfaces of roofs, ledges, and rafters where they like to build their nests. Once they have established themselves in an area, they and their offspring tend to return to the same area. Their nests are also a source of contamination from filth associated with nest material such as feathers and straw. If nests are found, they must be removed. Exclusion of birds includes the use of various IPM control measures such as habitat elimination, exclusion, and sanitation as well as chemical treatments.

Pest Control Measures

Pest Control Program

One of the first steps that food plant owners and managers can take to control pests is to have a written pest control program. This program outlines minimum guidelines for implementation of integrated pest management strategies as a means of prevention. Remove or neutralize all pest harborage, attractants, and breeding places within all buildings and property so that facilities and property may be pest free. Any signs or evidence of pests, whether young or mature, in or on all buildings or property and regardless of proximity to food products and processing will be considered a potential for contamination. A well-written pest control program must include, at a minimum, the following elements.

Management commitment and responsibility: The responsibility for implementation and enforcement of the pest control program will lie with the local plant management. Plant management will identify an individual, by name, who will be the specific employee to coordinate the local pest control program. This individual should maintain a valid pest control license. All plant personnel are to be trained to observe, note, and report all signs of pests. Plant Quality Assurance personnel are to monitor, advise, and guide management as to pest control practices and findings. The plant may want to identify other departments, such as sanitation or maintenance, and their specific responsibilities.

Definition of pests: Pests are defined for purposes of this policy, but not limited to, all rats, mice, and rodents, all insects, and all birds and vermin. Signs of pest activity are defined for purposes of this policy, but no limited to, live pests, dead pests, rodent droppings or urine trails, footprints or tracks, nests or evidence of roosting, and evidence of feeding or gnawing. Again, the plant may want to include pests specific to their region.

Inspection and reports: The plant plan will identify facility and grounds, monitoring them on a predetermined basis, ideally weekly. The inspection is intended to identify evidence of pest activity as described above. Inspection will include the outside of the facility, including the roof, the inside of the facility, and all pest control devices. Sticky boards, bait stations and catchall traps, and insect electrocutors are to be inspected twice monthly at a minimum. A record of findings will be prepared and submitted to plant management that identifies all findings of facility condition or pest activity that require corrective action. Signs of pest activity are to be entered in a pest-sighting log; a permanent record of all pest sightings. The findings report and sighting log shall include actions taken to correct findings from the prior report and their effectiveness. If a pest control operator is utilized, they will check with the sighting log during every service call.

Map of all devices: Prepare a plant diagram showing the entire perimeter of the plant and all walls and doorways. Mark on the diagram the exact location of insect light traps (electrocutors), bait stations, pheromone traps, catchall traps, and bird control devices. Each location will have a unique location number and identify the type of control device. The location number will be referenced in the plant pest inspection report.

Treatment materials: If plant employees apply treatments, they will be trained and licensed in application in accordance with state or local regulations. Samples labels and Material Safety Data Sheets (MSDSs) of any chemicals or baits applied will be maintained as part of the program.

It may be the choice of a plant or a company to utilize the services of a licensed pest control operator (PCO), someone with specific training in the identification, prevention, and treatment of pests. If the decision is made to use a PCO, the company or plant will want to make certain that they have a total program that meets the following minimum requirements. The choice of PCO should be made by local plant management, with the assistance of the QA Manager to verify credentials and ensure that there will be a good working relationship. The PCO will normally provide the company with a written service agreement or binder and a listing of services provided along with a current copy of the PCO license. The service agreement should include a list of all materials to be used by the PCO, and they must comply with USDA/FDA regulations. Chemical labels and MSDSs are to be submitted to the plant and maintained in a binder with all pertinent rodent program data and reports. The PCO will perform all services as per the applicable federal, state, and local laws and regulations.

Following each visit, the PCO will provide the plant with a detailed written report [8] that will include, but not be limited to

1. All services performed during the visit.
2. All locations where the service was performed by their specific identification. Document all chemical applications on an Insecticide Usage Log as shown in Table 10.1. Information on the log will include the specific treatment applied (i.e., insecticide, rodenticide, avicide), the target pest, the quantity applied, the location where applied, the method of application and dosage, and the date and time applied. The document will be signed by the PCO upon completion of application.
3. All findings of activity and types of pest activity found and the specific location where they were found. In addition to the above, the PCO will note on the report (extra pages may be used as needed) all conditions and areas to be corrected by the company to ensure pest-free facilities. The PCO will list on all reports any practices observed that encourage pest entry, harborage, or attraction. This will include all possible pest entry points in all buildings, all possible pest attractants on all plant property, and all possible pest harborage, both within buildings.
4. Date of inspection and service, including exact time in and out.
5. Name and signature of the individual making the inspection.
6. Physical condition of bait stations, traps, or insect electrocution devices

> All items in number 3 above will be listed on all subsequent reports until the problem is corrected or resolved with local plant management and the solution is noted on the PCO report.

The PCO will prepare the report before he leaves the facility and distribute copies at this time. At the completion of each service call, the report for that service will be reviewed by the PCO with the plant manager or QA representative. Following

Table 10.1 Insecticide Usage Log

			Cleo's Foods Insecticide Usage Log				
		Applicator Name:	Applicator Address:		Certification No.:		
Name of Insecticide and EPA Registration	Target Organism	Quantity Used	Where Applied	Method of Application	Rate of Application	Date Treated	Applicator Signature

Confidential Commercial Information.

this review, the plant manager or QA representative will sign the report, and a copy of all service reports will be left with the plant manager or QA representative at the time of the service call. Effective communication with the PCO will lead to a clear understanding of plant needs to facilitate a pest-free environment. A copy of all service reports will be sent to personnel at the location that the plant manager may designate for follow-up or corrective action. Ideally, the PCO will provide a quarterly trend report showing the frequency of service of bait stations or catchall traps. This information will be used to determine areas that require greater attention for control activities. Service reports should be maintained by the facility for one year.

Habitat Elimination

The first line of defense against pests is the perimeter of the plant. Elimination of possible sources of pest harborage outside the plant reduces the possibility that they will be available to get inside the plant. This means the creating an environment that is not hospitable to pests as well as creating a barrier that keeps them away from the plant.

Storage: Plants may have areas outside the plant where surplus equipment, empty ingredient containers, or pallets are held. These areas are often referred to as the "bone yard." The bone yard area must be maintained in a condition that will not provide harborage to pests. All machinery and other equipment, if stored outside, shall be stored in such a manner so as not to be or become a pest attractant or harborage. They will be kept off the ground by placement on racks or pallets to facilitate visual inspection for signs of nesting. Pipes or conduit will be capped at the end so that they cannot be used for nesting. This area is to be inspected periodically to verify that there is no harborage (see Figure 10.3).

All empty food containers stored outside, such as cans, buckets, drums, etc., will be placed in a pest-proof debris bin (must be food-grade or stainless) or will be washed in such a manner as to remove all food material both inside and out. All pallets, regardless of condition, that are stacked outside will be stacked on paved or concrete surfaces. In addition, they will be stacked no less than 6 feet from open soil. When pallets drums, buckets, etc., are stacked or stored outside, they will be stacked in such a manner as to minimize the harborage and attractant potential. As a general rule, this will be in widths not to exceed 5 feet; lengths not to exceed 10 feet; and with 2 feet spaces in between.

Foliage: All plant growth on the property shall be cut, pruned, and trimmed to minimize pest harborage. Weeds will be chemically treated or removed. Plant growth will be treated, as needed, with insecticides to ensure they are as close to pest free as possible. If there are other buildings or grounds surrounding the plant, they must be well maintained to prevent pest harborage or breeding sources. If they are not well maintained, consider a physical barrier, such as a solid fence, to prevent pest access.

Figure 10.3   Though stronger caps are preferred, taping will suffice to keep pests from nesting in pipes.

Trash containers: Outside of the building, eliminate food sources from trash cans or dumpsters. They must be pest tight bins or cans, have less than 1/16 inches opening when closed, and must remain closed. Trash carts, dumpsters, and inedible bins, staged outside the facility for disposal, will be adequately covered to prevent bird attraction (see Figure 10.4). Dumpsters may be uncovered; however, they must be emptied every 24 hours while processing and must be cleaned after every processing day.

Standing water: Make sure that plant grounds are well drained to prevent stagnant water or pooling water sources on grounds or on flat roofs that can become harborage areas. For mosquito control, eliminate standing, stagnant water, stock fishponds with mosquito-eating species, and use mosquito bacterial agents that won't effect environment or aquatic life.

Bird nesting and roosting: Know the habits of local birds, and modify the external habitat to prevent bird roosting. For example, sparrows like to sit on ledges, signs, and light fixtures to roost. Use of netting can prevent roosting on beams and rafters as a low-cost and long-term solution in large areas. Monofilament and steel lines can also be used to prevent roosting on ledges; use coiled wires or spikes. Barrier strips and wires are used to prevent roosting on beams, pipes, ledges, or light fixtures (see Figure 10.5).

Most of these don't shock; however, some provide a mild electrical current. Gels and pastes applied to flat surfaces where birds roost act as a repellent because birds

Figure 10.4 This outdoor trash can is not covered sufficiently to prevent attracting of pests.

Figure 10.5 Wires on light posts will prevent birds from roosting.

don't like a sticky surface when they step on it. These materials are nontoxic and are diminished by dirt, debris, and wet weather and will have to be replaced frequently. Sound devices or visual repellants (laser lights or predator models) work temporarily, but birds will eventually get used to them and their effectiveness decreases.

Exclusion

One of the primary means by which sanitary design works is to exclude pests, denying them access to food, warmth, moisture, and shelter. In other words, the objective of sanitary design is to keep pests out of the facility, eliminate the ones that get in, prevent damage by the ones that got in, and minimize their movement through the plant. The exterior of the plant is the first line of defense and should be evaluated to prevent the attraction of pests to the facility and entry into the facility. All buildings that are used for product ingredients, finished product, and food contact packaging will be pest proof in construction and operations to prevent pest entry and harborage.

Lighting: Outside the plant, select sodium vapor lights instead of mercury vapor as the former tend to attract fewer flying insects due to the light's wavelength. Move lights away from the building, and aim them to illuminate the building rather than having the lights on the building as that will then attract insects to the building [8].

Rodent runs: Gravel strips and runways around the building eliminate harborage and runways, and make for open areas that rodents don't like. A 30 inch wide and 4 inch deep trench with pea gravel around the plant perimeter will deter rodent burrowing, or flat cement shelf 24 inches belowground when a new foundation is poured (see Figure 10.6). Rodents can't burrow through gravel because it caves in behind them and they cannot chew cement.

Plant roof: Avoid stone ballast on the plant roof as this can hold contaminants. It is best to use single-membrane rubberized roofing as identified in Chapter 6.

Pest proofing: Seal all wall penetrations to prevent insect or rodent access. This includes all openings inside and outside the facility (e.g., cracks, joints, open conduit, or utility lines). As a general rule, openings that are in excess of 1/8 inch or larger at the smallest dimension constitute harborage and will be sealed or covered with pest-proof material. Key sizes to remember for openings are the following: an opening 1/2 inch square or round and smaller is rat and large bird proof, an opening 1/4 inch square or round and smaller is mouse and small bird proof, and screening for windows and doors 22 mesh and finer is fly and insect proof. In order to ensure that pest access is prevented at loading docks, load-leveler pits must be sealed and free of debris.

Bird proofing: Take measures to prevent bird nesting and roosting on or around facilities. Humane measures of preventing roosting and nesting, in compliance with state and federal environmental guidelines, should be incorporated in the plant IPM plan. Measures of bird proofing may include screening (approximately

Figure 10.6 A gravel strip around the plant eliminates a path for rodents, prevents them from burrowing, and provides a location for easy access to bait boxes.

¼ inch square), predator bird models (owls or raptors), or chemical controls (i.e., Avitrol as applied by a PCO).

Door and window control: The most common and easiest access for pests is through open or poorly sealed doors and windows. As a general rule, to be pest proof, doors and windows will remain closed and will be sealed so that no light is visible around them. If doors are to remain open, they should be effectively screened to prevent pest access and for plant security as indicated in the 2002 FSIS document entitled "FSIS Security Guidelines for Food Processors" [10]. Use plastic strips or air curtains for those doors that must be opened for any period of time. If plastic pest-proof strips are used, they will be clean and functioning. Shipping and receiving dock doors must be tightly sealed and closed. Otherwise, hang the plastic strips at loading docks to prevent pest entry or use a hood to form a seal around trailers. If an air curtain is used, the area must be under positive pressure with the air blowing toward the outside or the insects may be sucked in. Air curtains require a velocity of 1600 feet/min (approximately 18 mph) to be effective (see Figure 10.7). To be most effective, air curtains will be equipped with auto switches so that they will be on at all times when the door or window is open [7].

Inside the plant: To prevent accumulation of food or harborage of insects, caulk seal cracks at floor–wall junctions or at the seams of walls. It is important to repair leaky faucets, and squeegee floors after sanitation or in wet operations to prevent standing water that will attract pests. False ceilings are not recommended

Figure 10.7   This plant employs an automatic closing door, an overhead air curtain, and plastic strips to prevent pest entry.

because of the potential for accumulation of dirt and dust and for pest access and harborage. Air makeup systems pulling air in from the outside must be filtered at dimensions that will not allow entry by insects or other pests.

Sanitation

Poor internal housekeeping or outdoor sanitation can lead to attraction and breeding of pests that can eventually make their way into the manufacturing facility. Regular cleaning will prevent the attraction, breeding, or infestation of pests.

Externally

Maintain the grounds of the facility in a clean condition. Trash and refuse will be maintained at least 50 feet from the building and in sanitary conditions. All spilled product within 50 feet of the exterior of all buildings will be cleaned and removed immediately, and these areas are to be hosed and scrubbed once every processing or operating day. All spilled product more than 50 feet from the exterior of each building will be cleaned as needed. It is recommended that all such paved areas on property be hosed every week. In animal slaughter operations, it will be necessary to clean holding pens and chutes or live hang areas to remove all animal waste. Offal areas that open to the outside will require regular cleaning to eliminate materials that will attract pests. Daily removal of trash and offal will prevent attraction or harborage of pests. Clean up spills around dumpsters as well as storage tanks (flour, corn, syrup, oil).

Internally

Inside of all buildings, it is important to eliminate sources of food to starve out pests. This means that trash containers will be emptied no fewer than once every 24 hours and must be clean, inside and out of a production area. They may be uncovered if they are emptied and scrubbed once every 24 hours. Scrap and inedible containers will be emptied of all foodstuffs as frequently as necessary to prevent pest bird attraction. All area around garbage or debris bins must be cleaned every day. All spilled product and food items within all buildings will be cleaned and removed as quickly as possible. Overhead beams and structures, especially in dry or milling operations and seasonal operations, must be kept clean or they may become an attractant. Food will not be allowed to be stored or consumed in locker rooms, and lockers will be inspected as described in Chapter 9. Food or beverages will not be allowed in the plant on non-production days or during maintenance shifts, again as described in Chapter 9.

Warehousing

Visually inspect all incoming ingredients and packaging for signs of infestation, and reject any that are not pest free. Inspect the condition of delivery trailers for signs of pests, and reject trailers if they show signs of pest activity. Dock areas must be kept clean inside and out, especially around load levelers. All non-refrigerated storage areas shall have a rodent control strip. This strip shall be no less than 18 inches on the floor and 18 inches on the adjacent wall. This strip is to be painted white along the exterior walls only. Every two weeks, this strip must be inspected with a black light to detect the presence of rodent droppings and other signs of pests. Maintain 18 inch perimeters around storage racks or floor storage for cleaning and observation [8]. Rotation of ingredients (FIFU) is very important, especially in dry areas [14]. Re-inspect dry goods for insect or rodent activity if they are stored for periods of time longer than 3 months. Address product spills immediately in storage areas to prevent accumulation and pest attraction. All damaged or returned product shall be inspected upon receipt for indications of pest activity and will be stored under pest proof conditions until it is dispositioned.

Extermination

Insect traps: There are several different ways of capturing and eliminating insects from plant areas. These include light traps, such as insect electrocutors, stealth or pheromone traps for flying insects, and sticky boards for crawling insects. Stealth or pheromone insect units are preferred because they do not electrocute flying insects. However, insect electrocutors may be used in the facility but are not allowed in ingredient or product areas. Electrocution units rely on the use of ultraviolet (UV) light sources to attract flying insects. As they fly toward the light and into the unit, they come in contact with an electric grid and are "zapped." Because of this, electrocution units are to be kept away from stored ingredients or out of pathways for ingredients and in-process materials because insects coming in contact with the electrical grid may "explode," resulting in parts flying in several directions. Insect light grill openings must be large enough to provide light output to attract flies but not too open to create a personnel safety hazard. It will be necessary to change UV lights annually in spring as they lose effectiveness over time. Other types of traps only stun the flying insect and they fall onto sticky board, and some have no electricity but use light to attract insects to a sticky board (see Figure 10.8). For maximum effectiveness, all units using UV light for attraction are to be placed with the lights away from doors or windows, as the lights will attract insects toward the building.

Light traps must be away from competing light source such as sun or security lights to be effective and near where flies congregate. Bulbs are to be checked frequently and changed as needed to maintain effectiveness. Place these devices on the plant Master Sanitation Schedule so that catch pans are emptied frequently or sticky boards replaced to prevent accumulation or overflow of dead flies. In general, the cleaning frequency is weekly in peak seasons, and monthly at cooler times of the year when there is less flying insect activity. Light traps attract both male and female Indian meal moths, whereas pheromone traps attract males, relying on female scent for attraction [9]. Reduction of the male population by trapping will reduce mating and production of young insects. Use pheromone traps around dry ingredients where moths are more often a problem. Whichever type of trap is used, it is a good idea to monitor the contents for the amount of activity and the type of insects captured. This will give you a good idea of the effectiveness of your exclusion program and the types of control measures to apply. Keep in mind that insect light traps will only eliminate the adult population and will not control larvae or pupae. Also remember that different flying insects fly at different heights. For example, houseflies typically travel at 3 to 6 feet, whereas moths fly at 8 to 10 feet [1]. For that reason, it may be necessary to vary the height of your light traps depending on the type of flying insects you need to control. In addition, they are best located approximately 12–15 feet from doors or entrances to give flies an opportunity to get the light into their line of sight. Avoid using hanging flypaper because of exposed insect carcasses.

Sticky traps can be used on the ground to catch crawling insects. Like rodent catchall traps, they must be protected from damage and must be placed along floor–wall junctions, and along the typical paths of insects. For protection, they

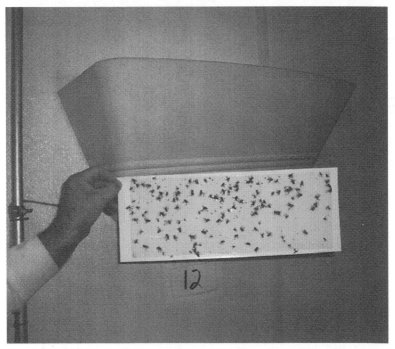

Figure 10.8 This light trap that does not electrocute but relies on fly attraction to the light, and subsequent capture on a sticky pad. This pad shows how effectively flying insects can be caught.

Figure 10.9 Placement of the bait station will be against the perimeter wall with the openings close to the path of rodent travel.

may be placed inside a piece of PVC with screws inserted to keep the PVC in place and to prevent it from rolling. If sticky boards or traps are used, they must be checked on a regular basis and replaced as they become filled.

Rodent control: Placement of rodent bait traps along the perimeter of the plant property as well as along the immediate exterior of the facility is a step that provides for exclusion as well as extermination. The intent is to provide bait to kill those trying to get onto the property or those that make it past the perimeter. External bait stations will be placed approximately every 30–50 feet [7]. They should be made of a material that will withstand varying weather conditions, have locked lids to prevent tampering, and be secured to prevent removal. Rodent bait should not be used inside the plant; however, there are nontoxic feeding blocks that can be used to monitor rodent activity if rodents are suspected inside the plant (see Figure 10.9).

If rodents do get into the plant, traps will be positioned between 20 and 40 feet apart and on either side of doors leading to the outside, to catch them before they can get into ingredients or product. Since rodents tend to travel along walls, this is where traps should be located. The trap must be positioned so that the opening is close to the wall giving the rodent easy access to be captured. They must be protected from damage by pallets, forklifts, or other equipment.

All bait stations and spring-loaded traps will be numbered and located on the plant map (see Figure 10.10). These devices will be inspected, serviced, and cleaned on a regular basis, and the person conducting the service will date and initial the

Figure 10.10 Proper positioning of spring-loaded rodent traps, on either sides of a door. They are protected from damage by pallets or forklifts with metal covers.

service record that will be maintained *inside* the device. Sticky boards may also be used for trapping of mice just as they are for trapping insects. Snap traps, however, are not recommended as they are not contained and may result in spillage of blood or other bodily fluids when rodents are caught. Keep in mind that mice can transmit hantavirus, and rodents carry other parasites and disease, so it is vital to handle them with care when removing them from catchall traps. Carefully remove dead rodents from traps, using protective equipment such as gloves to prevent virus spread, dispose of the carcass properly, and wash hands thoroughly afterward. Some plants have noted that feral cats on the premises help to control the numbers of rodents; however, cats can be sources of disease and should not be a considered part of the plant pest control program (see Figure 10.11).

Bird Control

The most effective means of bird control is through the elimination of their nesting areas and exclusion from the plant. Sanitation of outside areas and the removal of trash will eliminate their food sources. Revolving yellow lights bother bird's eyes, but they are most effective indoors and in dark areas [8]. Another method is the use of sound devices or predator models. The problem with these is that the birds tend to become accustomed to them, and eventually they will not stop having any effect.

Figure 10.11   Cats are not part of the IPM program. This cat would not be an effective mouser!

Chemical Interventions

Insecticides and Alternatives

As stated previously, one of the intentions of integrated pest management is to use methods that will reduce the need for chemical application. As much as possible, chemical use should be a last resort or a minor supplement to the comprehensive plant program [1]. Federal Insecticide, Fungicide and Rodenticide Act (FIFRA) is administered by the Office of Insecticide Programs under the Environmental Protection Act (EPA). This regulates the registration and use of insecticides. If they are used, they must be registered chemical compounds and must be used only for the intended purpose in compliance with EPA regulations. Plants are to have on file a copy of all labels for insecticides, rodenticides, or avicides used. It is best for insecticides not to be stored on the plant property; however, if chemicals are stored on site, they must be secured and held at proper temperatures away from food items. Insecticide containers must be fully and accurately labeled. Disposal of insecticides must be consistent with label directions and regulations.

Residuals and Nonresidual Insecticides

Some insecticides contain harmful chemicals, while others become inert on contact or within a brief amount of time. Residual chemicals kill both immediately and over time. Because they continue to act, they must only be used outside of the food facility in areas where plant employees will not be traveling, so they do not track pesticides into the food manufacturing facility. Their use is restricted by federal regulation. Included in this group are Baygon, malathion, Carbaryl (Sevin), chloropyrifos (Durisban), diazinon, and lindane. Nonresiduals are synthetic pyrethrins or Dichlorvous (DDVP) with a synergizer, and they degrade readily [5]. Pyrethrum only works on contact and needs to be applied in a small space (i.e., trash room or vestibule). These do not leave a residual kill and do not kill over time. There should be separate applicators, clearly identified, for residual and nonresidual chemicals.

Insecticide Application

Spray or fog application of insecticides is best conducted only by a licensed pest control operator. If a PCO is not used to apply chemicals, the employee performing the application must be fully trained and licensed where required by state law. There are three primary means of insecticide application; crack and crevice treatment, fogging, and fumigation. Most often, residuals are used for crack and crevice treatment, especially where insects are suspected or known to be colonizing. Because residual insecticides are effective for several days, crack and crevice treatment will be followed by sealing of the treated crack and cleaning of the entire treated area. Though residuals are not recommended for fogging, they may be applied to larger areas such as floors at wall junctions where insects are tracking or outside the plant to treat walls or floors in dumpster areas where flies might congregate or rest. With any residual application, all regulations for use must be followed for safety reasons.

Nonresidual insecticides are most often used for fogging or space spraying to knock down adults. Nonresidual sprays may be used in production or personnel welfare areas provided there are no ingredients, food products, or employee uniforms present. If nonresiduals are used in food contact areas, food should be removed and contact surfaces covered and cleaned following use. If fogging with nonresiduals, follow with a clean and sanitize procedure of all work areas, lockers, and equipment.

Fumigation is used in whole plant or sealed trailer application, typically with dry ingredients or grains, to destroy infestations by stored product pests. The material most commonly used had been methyl bromide; however, it was phased out in 2010. Alternatives include phosphine gas, aluminum phosphide, and magnesium. Studies are being conducted on the addition of carbon dioxide to these gases to improve efficiency and reduce dosage. Fumigation is a very technical and complex process that is best done by a trained professional. This individual will evaluate the material or area to be fumigated, calculate the gas dosage required, and have the needed containment, application, and protective material.

Battery-operated or plug-in auto mist machines not very effective in large areas or where air recirculation is heavy. They should be avoided in bathrooms or office areas as they may create discomfort to employees. It is recommended that plants alternate poisons to prevent pest tolerance. Fly bait may be used in dumpster areas if this tends to be an area of flying insect feeding and congregating during operations. If fly bait is used outdoors, it must be reapplied after a rain as it will be diluted and washed away. In all instances where chemicals or insecticides are used, personal protective equipment (PPE), such as gloves, mask, goggles, respirator, boots, etc., must be worn to protect the applicator from chemical exposure.

Alternatives to Chemical Insecticides

An alternative to chemical use is heat treatment, known as hypothermic treatment. In this process, the room temperature is raised to 126°F to 131°F in increments of 5°F to 10°F per hour. Though this treatment may take 8–30 hours, there is a 100% kill at all insect life stages, and death comes from dehydration and destruction of key enzymes and proteins [11]. In some plants, edible oil is used for spot treatment to knock down flies that get into storage or production areas. This is quick and efficient, does not require chemicals or cleaning, and is not as messy as using a fly swatter!

Drain flies or common nuisance flies, fruit flies, phorid flies, and moth flies as well as their larvae thrive on organic buildup in drains, sewage filters, p-traps and disposals (decaying organic material), which can then become breeding grounds. In addition to treating with sanitation, there are environmentally friendly materials that break up organic material in drains, eliminating food sources and helping to promote effective drainage.

Rodenticides

Rodenticides are poisons that typically thin the blood of rodents, resulting in internal bleeding and eventually death. Their use is permitted with precautions and instructions to prevent food contamination. One of the advantages of rodenticides is that the mouse or rat feeds on the bait and then returns to its nesting area, where it dies. One of the disadvantages is that rats and mice feed differently. Rats will gnaw and feed for longer periods, whereas mice tend to nibble and may not get a fatal dose. Also, rodents tend to be suspicious of changes in their surroundings, including the introduction of food. Introduction of bait stations may result in suspicion and the rodent only taking small, less than lethal amounts of bait at the initial feeding. If the rodent becomes ill but does not die, this may cause even greater avoidance of the poison bait. For this reason, it may be necessary to use a nonpoisonous bait initially until the rodent gets more comfortable with feeding and then switch to a poison bait. Bait stations should only be placed externally and must be sealed, numbered, and anchored in place to prevent removal by animals and unauthorized personnel [12]. Outside, rodent bait stations will be between 50 and 100 feet apart [7]. Rodenticides should not be used inside the facility unless as a last resort. If used inside, it will be placed in secure stations, and all human food and ingredient will be removed for the day. Bait will be removed before start-up and accounted for. In all cases, a log of bait feeding or replacement frequency will be made to track activity and identify problem areas that may require additional control measures.

Avicides

Bird control must not expose protected species under the Endangered Species Act of 1973. A special permit may be required for control methods that will kill birds. The Endangered Species Act (ESA) identifies those species that are in danger of extinction, and therefore protected. ESA regulates the use of insecticides where endangered or threatened species (those likely to become endangered) are concerned. Avitrol is one of the more commonly used bird control chemicals. It is mixed in with feed and usually doesn't kill the bird. It causes the bird to act in a very unusual manner that causes other birds to become scared and leave the feeding area.

What Do You Do if a Bird Gets in the Plant?

If your pest prevention methods are effective, then this should not be a problem; however, the likelihood is that eventually a bird will make its way into storage or production areas, so it is a good idea to have a plan to address capture and removal. There may be a temptation to shoot the bird; however, shooting birds is dangerous and may be unwise, especially if the bird is a protected species. In the event that a bird enters the facility, the bird will be isolated to one area (through door closure, etc.). If it is in a production area, all production in that area will cease. All exposed product and packaging will be covered. The bird will be captured, and trapping may be the best option unless the bird can be driven out of the room or out of the plant by the use of water hoses. Any equipment and personnel contaminated by bird droppings will be washed and sanitized. All contaminated product and packaging will be destroyed.

Preventing pests from entering the plant and infesting ingredients, product, or the facility requires the application of multiple strategies starting with a written pest control program and including habitat elimination, exclusion, sanitation, and elimination. If all elements are incorporated, the plant and the company can avoid product contamination that may lead to expensive losses of product or business.

References

1. Hedges, Stoy A. B. C. E., *Field Guide for the Management of Structure Infesting Flies*, G.I.E., 1998, pp. 34–35.
2. Hedges, Stoy A. B. C. E. and Dr. Mark Lacey, *Field Guide for the Management of Structure Infesting Beetles*, Franzak & Foster Co., 1996, pp. 13, 38–45.
3. Kramer, Richard Ph.D., *Bird Management Field Guide*, G.I.E, 1999, pp. 6–9.

4. Gould, Wilbur A. Ph.D., *CGMP'S/Food Plant Sanitation*, CTI Publications, 1994, chap. 11–12.
5. Troller, John A., *Sanitation in Food Processing*, Academic Press, 1983, pp. 198–199.
6. Gould, Wilbur A. Ph.D., *Total Quality Assurance*, CTI Publications, 1993.
7. Mannes, Cindy, Mark Sheperdigian, and James E. Sargent, Ph.D., Keys to Effective Pest Management in the Food Industry, *Food Safety Magazine*, The Target Group, Glendale, 2003.
8. Baumann, Greg, Pest Free by Design, *QA*, 2005, pp. 40–46.
9. Meek, Frank, Make "Green" a Primary Color This Year, *Food Quality*, 2005, pp. 78–81.
10. Taylor, Darlene, Screening Out Pests, *QA*, 2004, p. 17.
11. Imholte, Thomas J. and Tammy K. Imholte-Tauscher, *Engineering for Food Safety and Sanitation*, Technical Institute of Food Safety, Medfield, 1999, pp. 303–304.
12. Katsuyama, Allen M., *Principles of Food Processing Sanitation*, Food Processors Institute, 1993, pp. 342–348.
13. West, Ben C. and Terry A. Messmer, *Commensual Rodents*, Utah Stare University Extension, 1998, pp. 1–3.
14. Anon, Back-To-Basics Pest Management Pays Dividends to Food Plants, *Food Safety Magazine*, The Target Group, Glendale, 2005, p. 42.

Chapter 11

Chemical and Physical Hazard Control

> Why are we worrying about allergens all of a sudden? We haven't killed anyone yet!

Anonymous Food Company Employee

Chemical Hazards

This statement was actually uttered by a food company employee who was upset that certain procedures were being put into place to manage the handling of allergens within the plant. Fortunately, this employee left the company, and attitudes like this are few and far between. Chemical contamination in the plant may occur from several sources, including the inappropriate use of lubricants, cleaning compounds, and sanitizers. For this reason, specific plant operating procedures will be developed to ensure that all chemicals are controlled and that all containers for cleaning and sanitizing compounds are clearly labeled and that instructions for use are present. However, the chemical hazard that has gaining increased recognition is that of food allergens. Food plants must recognize the ingredients that they use in their products that are considered allergens and control their handling, labeling, and sanitation. To do so, it is important to understand food allergy causative agents and food allergen management.

True Food Allergy

Food normally doesn't provoke a response from the human immune system, the body's defense against microbes and other threats to health. In food allergies, two parts of the immune response are involved, according to researchers at the National Institute of Allergy and Infectious Diseases. One is the production of an antibody called immunoglobulin E (IgE) that circulates in the blood. The other part is a type of cell called a mast cell [13]. Mast cells occur in all body tissues but especially in areas that are typical sites of allergic reactions, including the nose, throat, lungs, skin, and gastrointestinal tract.

People usually inherit the ability to form IgE against food. Those more likely to develop food allergies come from families in which allergies such as hay fever, asthma, or eczema are common. A predisposed person must first be exposed to a specific food before IgE is formed. As this food is digested for the first time, the body for some reason sees tiny protein fragments as harmful and prompts certain cells to produce specific IgE against that food. The IgE then attaches to the surface of mast cells. The next time the particular food is eaten, the protein interacts with the specific IgE on the mast cells and triggers the release of chemicals such as histamine in an effort to protect the body from the protein. It is this release that produces the symptoms of an allergic reaction [1]. If the mast cells release chemicals in the nose and throat, the allergic person may experience an itching tongue or mouth and may have trouble breathing or swallowing. If mast cells in the gastrointestinal tract are involved, the person may have diarrhea or abdominal pain. Skin mast cells can produce hives or intense itching. The most severe bodily reaction is anaphylaxis, a Greek word meaning protection. Anaphylaxis is a generalized shock reaction and may result in multiple symptom response and even systemic failure. Anaphylactic shock reaction can be fatal if not treated quickly. Allergen foods that can cause fatal reactions break down as follows: 60% peanuts, 30% tree nuts, and 10% milk and fish (see Figure 11.1).

True food allergy is caused by specific proteins in foods, such as gluten in wheat. The food protein responsible for an allergic reaction are heat and pH stable and are not eliminated by cooking or frying or by stomach acids or enzymes that digest food [10]. There are eight (8) primary food items associated with severe allergic reactions in a certain segment of the population, and they are listed in Table 11.1. It is estimated that about 90% of food reactions come from the Big 8 [2]. The items listed in the table below are the cause of most food allergic reactions in the United States; they must be managed in the manufacturing environment and clearly labeled on product packaging.

Food Intolerance

In addition to food allergy, there are other reactions to foods that are not immune mediated. Food intolerance is a metabolic disorder due to an enzymatic deficiency

Figure 11.1   **Examples of peanuts and tree nuts.**

Table 11.1   Big Eight Food Allergens

Allergen Food	Common Ingredient Names
Milk (dairy)	Casein, whey, butter, custard, nougat
Eggs	Albumin or albumen
Peanuts	
Tree nuts (Picture 9-1)[a]	Almonds, cashews, Brazil nuts, walnuts, pistachio, macadamia, pecans, hazelnuts (or filberts), pine nuts (or pinõnes)
Soybeans	Textured soy protein, textured vegetable protein
Wheat	Gluten
Fish	Cod, trout
Crustacea	Crab, shrimp, lobster, crawfish

[a] Each tree nut is a distinct allergen and must be treated as unique. (From Deibel, Dr. Virginia and Laura Berkner Murphy, Writing and Implementing an Allergen Control Plan, *Food Safety Magazine*, The Target Group, Glendale, 2003/2004.)

Note: In Europe, celery is also recognized as a food allergen. In Canada, sesame seed is recognized as a food allergen.

resulting in inability to metabolize certain materials [10]. Celiac disease or celiac sprue results from the inability to metabolize wheat gluten, the major protein of wheat. The harmful proteins that cause this are from primarily wheat, rye, barley, and oats. These proteins are not present in rice or corn, so foods made from these grains are safe for celiac patients. Celiac disease is generally not fatal; however, exposure to gluten causes damage to the cilia of the small intestine. This damage can result in malnutrition in the individual and in rare cases can result in intestinal tumors [11].

Another metabolic disorder is lactose intolerance, the inability to metabolize milk sugar, or lactose. Dairy products are associated with lactose intolerance, which typically causes intestinal discomfort, but is not life threatening. There are over-the-counter treatments for lactose intolerance; however, the best course of action for persons with lactose intolerance or celiac disease is to avoid the foods that cause problems [10].

A third area of food response for manufacturers to consider is food idiosyncrasies. These are responses to foods or food additives resulting from unknown mechanisms. Examples of idiosyncrasies are sulfite-induced asthma, monosodium glutamate (MSG) sensitivity, and hyperactivity from yellow food coloring [10]. Approximately 150,000 people in the United States are susceptible to asthmatic reactions to sulfites at a level greater than 10 ppm. Sulfiting agents are used in food products as a preservative of a bacterial inhibitor. For example, in wine making, sulfites are used to inhibit the growth of undesirable yeasts or other bacteria. If the sulfite level exceeds 10 ppm, the product label must bear the statement "Contains Sulfites."

MSG is added to foods to enhance flavors, and there is great debate over the true potential effects of MSG on people. Studies have proved inconclusive on the ability of MSG to produce reactions in "sensitive" people. The reaction in sensitive individuals is not life-threatening; however, they sometimes develop headache and flushing that gradually subsides over time. MSG must be labeled when it is present in foods because of the potential for consumer reaction and federal regulation. Yellow Dye #5 (Tartrazine) labeling is also required because of its perceived effects of hyperactivity in children. While several recent studies have raised questions as to the impact of Yellow Dye #5, federal regulation still requires labeling if it is present in food products. As with foods that result in true food allergies or metabolic reaction, the sensitive consumer is better off avoiding the food or products containing the food. However, manufacturers still have an obligation to produce products that will not expose consumers to sensitizing agents without their knowledge.

Regulatory Requirements

Over the past ten years, there has been increasing regulatory activity in the area of allergens. This may be partly due to the statistics surrounding allergens in foods.

It is estimated that approximately 7 million Americans suffer from food allergies. This statistic breaks down to about 2% of adults and 3% of children. In addition, there are almost 30,000 emergency room treatments and 150–200 deaths each year related to food allergens. As a result, there is an increasing awareness of allergen cross-contamination by industry and regulatory agencies. This awareness has increased the number of food recalls due to allergens from 0 recalls in 1988 to approximately 124 in 2003 (FDA). The reasons for the increase in recalls are production cross-contamination and mislabeling of the product by the producer. However, consumers also have indicated that they are sometimes confused by the use of ingredient names on food labels of which they are not familiar. The use of labeling statements such as "Free From" (i.e., Free From Soy), "May Contain" (i.e., May Contain Soy), or "Made in a plant/on a line that uses (allergen)" has actually caused confusion among consumers. Consumers say they don't find these statements beneficial. For these reasons, both primary food regulatory agencies, the FDA and USDA, have taken a more active role in food allergen management.

In 2004, both houses of Congress approved the Food Labeling and Consumer Protection Act, which went into effect in January 2006 and amends the Federal Food, Drug and Cosmetic Act (21 USC 343). The act imposes requirements on both food manufacturers and government agencies. Industry is obligated to provide clear and accurate ingredient labeling. The primary provisions of the act are to provide easy-to-understand labeling for consumers, analyze the means by which foods are unintentionally contaminated with allergens, advise industry on best practices to prevent cross-contamination, among other provisions to assist consumers and industry.

In July 2005, FSIS issued Notice 45-05 entitled "Verification of Activities Related to an Establishment's Controls for the Use of Ingredients of Public Health Concern." The notice gave guidance to inspection personnel for verifying plant procedures to control ingredients of public health concern that trigger allergy response or metabolic intolerance. USDA's rationale for the issuance of the notice was the increasing number of product recalls resulting from misbranding or undeclared allergen ingredients in meat and poultry products. The notice provides inspection personnel with brief information that they need to know regarding ingredients (The Big 8) that are known to cause immunologic and metabolic reactions in sensitive individuals. It also identifies ingredients that have potential for adverse reactions such as MSG, sulfites, and Yellow #5 [6]. For the purpose of identification in this section all of these ingredients will be referred to as "allergens."

FSIS Notice 45-05 instructs inspection personnel to evaluate plant HACCP plans, verify that the plant has addressed allergens in the flow diagram, and risk assessment. It further instructs them to verify that the establishment has adequately incorporated allergen management into its food safety systems either through inclusion in HACCP plans, SSOP, or prerequisite programs. If the plant uses a prerequisite program to control allergen ingredients, inspectors are to verify that an Enforcement, Investigations and Analysis Officer (EIAO) has assessed the establishment food safety system. If the plant has not addressed allergens in its control programs, has not followed its own procedures, or has mislabeled product, the inspector is to issue a Noncompliance Report [6].

The question that has arisen is how to address allergens in the HACCP plan. First, they may be addressed in a listing of ingredients used by the plant as shown in Table 11.2. Second, they can be identified in the process flow diagram at the receiving stage with identification where they are introduced in the production process. An example of this can be found in Table 11.3. Finally, they should be included in the Hazard Analysis for determining whether they are a hazard that is reasonably likely to occur, as demonstrated in Table 11.4.

Table 11.2 Cleo's Foods HACCP Plan, Fully Cooked Not Shelf-Stable Bologna, Cooked Salami, Luncheon Loaf Incoming Ingredients and Packaging

1. MEAT INGREDIENTS	Pork, Beef
2. NON-MEAT FOOD INGREDIENTS:	Salt[a] Dextrose Water[b]
SPICES/FLAVORINGS	Black Pepper Nutmeg Coriander White Pepper
RESTRICTED INGREDIENTS	Nitrite
ALLERGEN INGREDIENTS	Casein Soy
PRESERVATIVES AND ACIDIFIERS	Salt[1]
3. OTHER	Inedible Casing
4. DIRECT CONTACT PACKAGING MATERIAL	Plastic film

[a] Salt is listed under both non-meat food ingredients and preservatives and acidifiers.

[b] Water used for functional purposes during manufacturing process.

Management Signature _____ *Date* _____

Confidential Commercial Information

Table 11.3

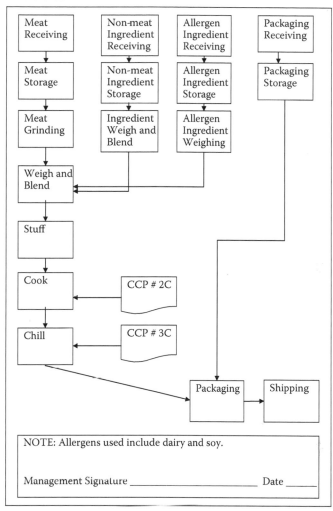

NOTE: Allergens used include dairy and soy.

Management Signature _____ Date _____

Allergen Management

From the scientific community, there is no defined or consensus threshold level because some individuals show symptoms at extremely low-level exposures. From a regulatory standpoint, neither USDA nor the FDA has set an acceptable limit for allergenic materials as there is no known limit to the amount of allergenic protein needed to elicit an allergenic response. The FDA's Threshold Working group is currently evaluating the potential of establishing a minimum level that may be applied to allergens in foods; however, many scientific, risk, analytical, and legal issues

Table 11.4 Cleo's Foods HACCP Plan, Fully Cooked Not Shelf-Stable Bologna, Cooked Salami, Luncheon Loaf, Hazard Analysis

Process Step	Food Safety Hazard	Reasonably Likely to Occur	Basis	If Yes in Column 3, what measures could be applied to prevent, eliminate, or reduce the hazard to an acceptable level?	Critical Control Point?
Meat Receiving	Biological Pathogens: *Listeria monocytogenes, Salmonella,* staph parasites; trichina worm	Yes No	Raw meat and poultry have been associated with pathogens. Plant uses Certified Trichina Free pork, COA.	There is a later step in the process designed to reduce or eliminate pathogens.	No
	Chemical: pesticides, drug residues	No	Supplier Letter of Guarantee		No
	Physical: Foreign material; broken needles, buckshot, hooks, knife blades, wood	No	No evidence of any historical occurrence at this location.		No

Non-meat Ingredient Receiving	Biological: None			
	Chemical: None			
	Physical: Foreign Material; glass, metal, wood	No	No evidence of any historical occurrence at this location. Visual inspection of incoming materials.	No
Allergen Ingredient Receiving	Biological: None	No		
	Chemical: Allergen	No	Supplier Letter of Guarantee. Plant Prerequisite Program.	No
	Physical: None	No		

must be resolved before this can be established [9]. For these reasons, it may be wise for a company to establish a zero tolerance for allergen cross-contamination and implement an allergen management program.

The purpose of the allergen management program is to employ multiple hurdle practices that will prevent the cross-contamination of products with allergens and ensure that products containing allergens are clearly labeled. There are several strategies to be employed by manufacturers; however, they can be broken down into four distinct areas within the manufacturing plant: Ingredients, Production, Packaging and Labeling, Sanitation, and Consumer Feedback.

Ingredients

One of the very first areas of allergen control and management is receiving and handling of ingredients. Companies should provide clear directions to their suppliers that they are not to change ingredients, especially in multiple component ingredients (i.e., spice blends) without adequate notification, especially if the change introduces an allergen. As a means of verification, it is wise to establish a process to evaluate labels of incoming raw materials against their material specification, especially with compound ingredients, to ensure that there have been no changes that have introduced an allergen. Auditing of the suppliers is also recommended to establish the adequacy of their allergen management program. If it is not possible to conduct audits of all suppliers due to numbers, cost, or location, it is reasonable to expect them to have a third-party audit by a reputable firm that includes allergen management in their audit process.

Another area of control is to provide clear and secure storage of allergenic food items. Do not store allergen-containing ingredients above non-allergen or different allergen ingredients (see Figure 11.2). If the plant does not have the space to provide adequate separation between pallets of allergen ingredients and non-allergen, or different allergen ingredients, it is helpful to shrink-wrap pallets of allergen ingredients to provide separation.

Clear identification of pallets of allergen ingredients with stickers helps plant personnel immediately recognize those ingredients with ingredients and provides easy visual verification when pallets are stored properly. This can be accomplished by applying stickers with "A" to identify the ingredients as allergens, or for greater clarity, by applying a sticker that identifies the specific allergen, as shown in Table 11.5 (see Figures 11.3 and 11.4).

Because of the nature of food manufacturing plants; ingredient storage on pallets and the use of forklifts and pallet jacks to transport ingredients through the plant, it is almost a given that there will be packaging tears and ingredient spills. When this happens it is vital as part of the plant operational sanitation program to have a defined cleanup and disposal procedure for damage to allergen ingredient packaging materials or spills at receipt, during storage and when transporting

Figure 11.2 **Allergen ingredients (wheat) improperly stored above non-allergen ingredients.**

Table 11.5 **Allergen Ingredient Identification**

Allergen	Sticker Identification
Soy	As
Wheat	Aw
Eggs	Ae
Dairy	Ad
Peanuts	Ap
Tree nuts	At
Fish	Af
Crustacea	Ac

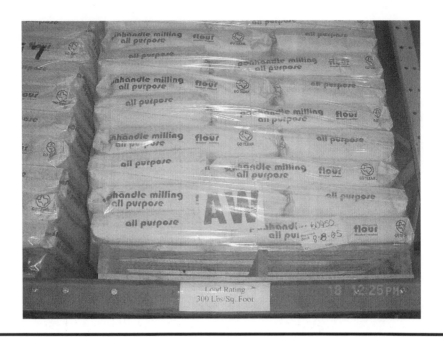

Figure 11.3   Illustrations of plant using stickers to identify specific allergen category (wheat on left, dairy on right) to aid in storage and handling.

Figure 11.4   Illustrations of plant using stickers to identify specific allergen category (wheat on left, dairy on right) to aid in storage and handling.

within the plant. This procedure will address the cleanup of the allergen ingredient that will remove the hazard without potential cross-contamination of other ingredients or in-process product.

Production

Food manufacturers have a responsibility to protect food allergen–sensitive consumers and must understand hazards inherent in their processes and how they can be prevented or minimized. For this reason, it is vital that the plant conduct an allergen risk assessment to determine where in the process allergen cross-contamination may occur. Some of the greatest challenges to the allergen management program occur in production. This is because there are many people involved, many ingredients that are brought together, and multiple processes that occur at the same time. It is important to address all of these items to control and prevent contamination.

1. People: Human error is a consideration that should be addressed through thorough training. All employees, including sanitation and maintenance staff, are to be trained regarding food allergens in general and should also be given training specific to their job function. They should be aware of the allergen ingredients used in the facility, trained how to handle them to prevent cross-contamination, and report incidents where cross-contamination has occurred. Maintenance will be made aware of allergens and their role in allergen management, such as employing sanitary design to prevent niches where allergen ingredients may become entrapped. Where possible, dedicate trained personnel, processing, and packaging lines to handle or process specific allergic ingredients [2].
2. Sanitary design: As described in Chapter 6, sanitary design of plant equipment is extremely important for the prevention of microbiological contamination. This is also true with prevention of allergen contamination. It is important to have equipment that is designed to avoid niches that can harbor ingredients or products that can result in cross-contamination. Equipment should be non-absorbent to prevent holding of allergen ingredients [4]. The equipment should be made for easy and thorough cleaning to remove all food residues. When designing manufacturing lines, it is important to avoid conveyors that cross over each other so that allergen materials from the upper line do not drop down and contaminate materials on lower lines.
3. Segregated equipment: Ingredient bins containing allergen ingredients and scoops or other utensils used for allergen ingredients will be clearly identified. They will be dedicated to the specific allergen ingredients and will not be shared with non-allergen ingredients [3].
4. Scheduling: If production lines are shared between products that do not contain allergens and those that do, it is preferable to run the products that

do not contain allergens first, immediately after the sanitation shift. Ideally, product scheduling is done so that allergens are run last on the lines each day, just prior to cleanup. Also, where possible, schedule long, continuous runs of products containing allergens in order to minimize changeovers [2]. Where this is not possible or where a changeover results in going from a formula with an allergen to a formula with a different allergen, initiate thorough cleanup between the products. Cleaning will vary, depending on the types of ingredients or finished products. In a dry product plant, such as cereal or crackers, wet cleaning is not recommended, and so a dry clean procedure will be used. This may involve scraping, brushing, or vacuuming to eliminate residue of allergens. In plants with "wet" products, such as meat and poultry, a wet cleaning can be done with soap and water to remove the protein residue. While there are some rapid test methods for certain allergens, the standard may have to be "visually clean" for equipment where there is no rapid test method to ensure that allergen residues are removed. A plant may also employ what is referred to as "push through" using inert materials (salt or flour) to flush the system between ingredients or blends. The push through material is then labeled to identify the specific allergen ingredient and stored for use in future production of like allergen blends [1].

5. WIP and Rework: Certainly, product formulas containing allergen ingredients will be clearly identified for plant employees, especially those handling ingredients or working in kitchens. It is also vital that those individuals handling work in process (WIP) materials and rework have a clear understanding of the allergens in the products they are handling. Food processing plants should develop a clear, sensible rework policy to prevent cross-contamination with allergens. The best advice is to use only like into like where rework is concerned [2]. One means to facilitate this is to create a rework grid, as illustrated in Table 11.6, to indicate which products may be used as rework into other products.

Table 11.6   Rework Usage Grid

Item Code	12345	67890	13579	24680
Contains	Wheat	Soy	Wheat	Wheat, Soy
Use In:				
12345	Yes	No	Yes	No
67890	No	Yes	No	No
13579	Yes	No	Yes	No
24680	Yes	Yes	Yes	Yes

Table 11.7 Sample label with the allergens listed parenthetically in the ingredient statement and with a "Contains" declaration at the end of the ingredient statement.

APPROX 120/1 oz. BY WEIGHT

105294 A   **Cleo's Foods**   KEEP FROZEN

CHICKEN POTSTICKERS

(CRESCENT SHAPED DUMPLINGS FILLED WITH CHICKEN & VEGETABLES)

**NO MONOSODIUM GLUTAMATE ADDED

FOR FOOD SAFETY, FOLLOW THESE COOKING INSTRUCTIONS. PAN-FRY METHOD: Place 2 tsp. Vegetable oil in a medium hot pan (preferably Teflon coated). Place frozen potstickers in pan, bottom side down. When bottom is golden brown, add 1/3 cup of water. Cover and steam for 3 minutes. * STEAM METHOD: Steam Frozen potstickers for: 6 - 7 minutes. * MICROWAVE METHOD: (1250 watts)* Place frozen potstickers on microwave safe dish with 2 tablespoons of water and cover with plastic wrap. Allow 40 seconds per potsticker. Cooking time may vary with equipment. * * Cook thoroughly, minimum internal temperature should be 165°F for at least 15 seconds.

INGREDIENTS: Bleached and Enriched Wheat Flour (Wheat Flour, Malted Barley Flour. Niacin, Reduced Iron, Thiamine Mononitrate [Vitamin B1], Riboflavin [Vitamin B2], Folic Acid), Cabbage, Dark Meat Chicken, Water, Soy Sauce (Water, Soya Beans, Salt, and Wheat Flour), Celery, Sugar, Soybean Oil, Green Onion, Garlic, Modified Food Starch, Sesame Seed Oil, Dehydrated Onion, Chicken Broth, Cottonseed Oil, Salt, Spice, Sodium Benzoate.

**Beyond the small amount naturally occurring in soy sauce.

CONTAINS: Wheat, and Soy.

NET WT. 7 LBS. 8 OZ. (3.40 KG)

Cleo's Foods
www.cleosfoods.com

It is also recommended that rework be labeled to identify the specific allergens present and further distinguish how it can be used. Work in process materials are those items that may not go into finished packaging for various reasons but are stored until ready to be packaged. Whether they are stored on racks, in vats, or in boxes, it is highly recommended that they be labeled as to the item and contents, especially the allergen ingredients, so that they are not mislabeled when they are brought into the production area for packaging.

6. Third-party auditing: This is an extremely valuable tool that is used to validate the entire allergen management program. The use of a third-party auditor, especially one who is experienced in allergen management, as well as plant manufacturing processes, can provide valuable insight into the construction and implementation of an effective allergen management program [3].

Packaging and Labeling

Finished product labels are required by federal regulation to disclose all allergen ingredients in the finished product in the ingredient statement. As indicated previously, allergen ingredients will be listed by their common names either parenthetically [i.e., Whey (Milk)] within the ingredient statement or at the end of the ingredient statement with a "contains statement" (i.e., Contains Milk) (Figure 11.7).

It is vital that labeling accuracy be verified during label preparation, at printing, and finally during packaging at the plant level. One way to do this is to have a trained individual prepare the ingredient statement for the label based on the formula provided by product development. This information is then submitted to a label printer, if labels are not printed in the plant, and the labels are prepared. Once the label has been printed, the plant can have the printer send an out of run sample to the plant for evaluation to ensure that the ingredient statement is accurate. Check formulas against packaging at least twice per year or when approved formula changes are made to ensure allergens are declared. Have a daily check procedure to ensure that the correct packaging is used on products containing allergens. Obsolete packaging that does not have product allergens listed cannot be used and should be placed on QA Hold and segregated for disposal [2].

Note: Some packaging materials contain wheat-based release agents. Determine from packaging suppliers if packaging contains wheat-based release agents. A letter indicating that no wheat-based release agent will be provided where none is used.

Sanitation and Allergen Control

As indicated earlier, there are several causes of allergen cross-contamination that can be controlled through an effective management program. Some of the product cross-contamination may be the result of inadequate cleaning of shared equipment that may be prevented by effective sanitation. Certainly, equipment design is critical, and equipment should be engineered for cleaning access. In addition, a well-designed sanitation program, written with allergen cleanup in mind, is important for prevention.

Sanitation plays a major role in addressing allergen cleanup based on the ingredient and the equipment surface. Where possible, conduct a wet cleanup between allergens and non-allergens. Wet cleaning is preferred as allergen proteins tend to be soluble in hot water and can be removed by detergents [1]. In plants where wet cleaning is not conducted, cleaning may involve scraping, brushing, or sweeping of dry allergen ingredients to remove them from the equipment and environment. Use of air hoses to clean equipment is not recommended as these tend to spread allergens by blowing them to different areas. Instead of using air hoses, consider using a vacuum for dry ingredient removal [1].

Some plants use a "visually clean" standard when conducting inspection cleanup between allergen ingredients. This means that if the surfaces look clean, they are considered clean. There are also test kits available to food manufacturers to verify cleaning. ELISA-based test kits are available for sanitation validation of some allergen compounds [1]. In addition, the plant can conduct a laboratory analysis of "first through" product. This means that as they change from a product containing an allergen to a product not containing an allergen, the first 1 to 2 pounds will be taken out of the product flow (i.e., out of a stuffer or a fill pipe). This quantity will be disposed of, and the subsequent material will be tested for the allergen ingredient in question. If the prior fill has done the job of flushing out the system, the subsequent filling will not test at a detectable level for the allergen ingredient.

Consumer Feedback

As consumer complaints are received, it is important to have a trained individual handle the calls and evaluate them for indications of problems with allergens. This person understands that consumers, especially those who have family members with food allergies, are getting more and more sophisticated regarding food manufacturing processes and product labeling. They want a considerate individual who can provide concise answers to questions and assistance with understanding products that they or their family can consume. This individual may also evaluate complaints and trend data to establish whether there are repeat problems or problems with similar root cause.

Disposition/Liability

Products suspected of containing undeclared allergens or cross-contaminated with allergenic ingredients are considered adulterated and should be rejected and placed on QA Hold. If those products containing undeclared allergens are in distribution, they are subject to Recall. The class of the Recall may be determined by the potential health risk associated with the allergen. The more severe the risk from the allergen, the higher the class of the recall. As indicated earlier, peanut allergen has been associated with fatal reactions in susceptible individuals; therefore, undeclared peanuts in a food product would result in a Class I recall as this is a situation in which there is a reasonable probability that the use of, or exposure to, a violative product will cause serious adverse health consequences or death.

Cross-contamination with wheat, which has not been identified with fatal reactions, would more likely fall into a Class II recall as this is a situation in which use of, or exposure to, a violative product may cause temporary or medically reversible adverse health consequences or where the probability of serious adverse health consequences is remote. It is not likely that undeclared allergens would fall into a Class III recall. Despite the recall class, no company wants their product involved in a recall due to the potential harm to consumers, the expense, and the loss of business associated with this action.

Manufacturers would also be well served to understand the liability they bear if their products result in injury or illness to consumers. Manufacturers are responsible for damages caused under strict liability laws. In certain instances, negligence does not have to be demonstrated. Most liability is under state rather than federal law, and liability law may vary from state to state. However, most states do allow consumers to sue manufacturers for damages resulting from injury/illness caused by food allergies. The plaintiff may sue under product liability theory; a variety of claims are possible, including negligence, strict liability, breach of warranty, failure to warn [12].

In the case of **negligence**, the plaintiff alleges that manufacturer negligence caused harm. For example, if the manufacturer uses the same equipment for allergen and non-allergen product and fails to clean between runs, the plaintiff may sue due to the manufacturer's failure to clean between allergen and non-allergen run even though they know that they should have [12].

A case of **strict liability** occurs where the manufacturer is held liable regardless of whether negligence on their part is established. The manufacturer is liable simply because the product in commerce caused harm [12].

In a **breach of warranty**, the plaintiff claims that the manufacturer warranted that the product would cause no harm when sold. If the product caused harm, then the manufacturer breached the warranty and is liable for damages [12].

Product mislabeling may result in a **failure to warn** claim. The plaintiff argues that the manufacturer knew or should have known of allergens in their product. The manufacturer had a duty to warn consumer of allergens in the product through a notice (i.e., labeling). The manufacturer is liable for failure to inform of the allergens. When the presence of allergen or sensitivity-producing food is obvious, the manufacturer does not need to provide a warning. Examples of situations where the presence of allergens would be obvious are a bag of peanuts or a carton of milk. The problem generally arises when the food item contains an allergen ingredient that is not reasonable to expect in the food, and the ingredient is not declared due to omission or cross-contamination.

Examples would include a restaurant that includes peanut butter in chili or where a company making a product enclosed in a flour wrap uses egg to seal the wrap but doesn't declare its presence [12].

In a liability claim, the plaintiff may bring all four causes of action into the suit when seeking damages for injuries for alleged "defects." The damages awarded to the plaintiff may be compensatory and/or punitive. Compensatory damages provide compensation for injuries and may include medical expenses, lost wages, "pain and suffering." The pain and suffering are typically three to five times the plaintiff's out-of-pocket expenses, but may be higher, depending on the findings of the court. Punitive damages are liability for company failure to warn if judged to be as a result of "actual malice" or "reckless disregard." In other words, the defendant was aware of the dangers posed but took no action to eliminate the hazard. In addition to damages paid, the company will suffer due to negative publicity and loss of customers. For these reasons and for protection of consumers, inclusion of allergen management into the food safety and sanitation is critical [12].

Physical Hazards

Physical hazards are hard objects of specific size that present potential for injury to consumers. The FDA defines physical hazards as hard objects between 7 and 25 mm, as this size is considered to be most likely to cause a dental injury or choking hazard. They are materials that are not part of the product or expected by the consumer to be found in the product. Examples of true foreign material that can constitute a physical hazard are wire, metal, glass, stones, plastic, and wood. By their nature and size, they present the opportunity for severe injury to consumer teeth and gums, they can become lodged in the throat causing choking, or they can penetrate the soft tissue of the esophagus, resulting in infection. Thus, these objects must be excluded from the food being manufactured. Other foreign objects such as paper, hair, or soft plastic do not pose as significant a hazard; however, all efforts must be taken to keep them from entering food products as they are certainly not desirable for consumers (see Figures 11.5 and 11.6).

Naturally occurring material, while still undesirable, is different from foreign material in that it may be associated with a food product or might be expected in product. Examples of naturally occurring materials are stems in spice, gristle in meat, or seeds in fruits. These items can become hazards, depending on their size and hardness; for example, olive or cherry pits can damage teeth, or bone in the

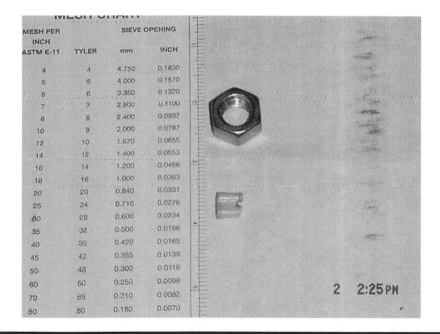

Figure 11.5   Hard foreign material such as metal nuts and hard plastic can result in injury to teeth or soft tissue.

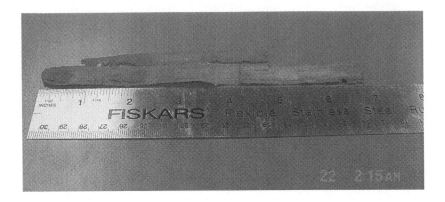

Figure 11.6 Soft foreign materials such as cardboard are less likely to cause injury but are undesirable for consumers.

range of 7–25 mm can injure the mouth or be a choking hazard. As such, it is a good idea to have an inspection program for these types of materials if you make them as finished products or use them as ingredients.

There are several measures that a plant can take to prevent product contamination with physical hazards. Use of bone collection devices on meat grinders can remove a significant portion of bone material and some of the connective tissue such as tendon (see Figure 11.7).

Figure 11.7 Bone collector on meat grinder.

Sifters on flour lines can remove some of the hardened material that may be associated with flour milling. Magnets are one of the oldest forms of methods for removing metal from foods and can be used to remove fine magnetic (i.e., ferrous) materials and some grades of stainless steel from of dry or free-flowing liquid ingredients [7]. GMPs, as described in Chapter 9, can prevent introduction of pens, jewelry, or hair to product, and operational sanitation inspection can identify opportunities for materials to enter product streams. Of course, metal detection is very common in the food industry and has application at the beginning of the line as well as at the end of the line. Poultry plants have begun using X-ray technology to detect bone in "boneless" products and X-ray equipment can be used to detect foreign materials in foods such as glass, metal (ferrous and non-ferrous), stones, hard dense plastic, and certain rubbers [7]. All of these devices require monitoring during production to establish that they are working properly and require cleaning to function effectively and not pose a microbiological hazard to product [5].

Sanitation Role in Physical Hazard Prevention

Physical contamination can also come in the form of hardened material such as dough or starch buildup and entry into food streams. Sanitation procedures, as identified in Chapter 7, will be designed to remove soils specific to the product and the potential physical hazards presented. It can come in the form of equipment parts, plastic or metal shavings, and welding spelter or wire that are left in product streams because of poor maintenance practices. As such, sanitation and maintenance both play a significant role in preventing foreign inclusions.

Detection equipment must meet the same sanitary design requirements as with all other food-handling equipment. This means that the equipment will be made of nonabsorbent material and have no cracks or seams and dead spaces. There should be no parts that might fall off into product during operation. The equipment will also be easily cleanable to a microbiological level and provide access for cleaning and inspection.

Sanitation procedures for detection devices will vary based on the type of equipment and the environment. For example, magnets used in a dry flour or spice process will likely require brushing or vacuuming but not wet cleaning. In fact, cleaning of magnets will depend on the magnet type and whether they are provided with mesh sleeves that are removed for cleaning while drum magnets are often self-cleaning as they rotate [7]. Other devices, such as metal detectors, may be able to withstand wet cleanup and sanitizing, provided low pressure is used to prevent water from entering the detector casing. In all cases, it is advisable to consult the equipment manufacturer for any specific cleaning requirements or restrictions. It is important to know whether the equipment requires lock out/tag out, protection from liquid, requires only wipe down cleaning or can withstand standard wet cleaning.

Finally, the sanitation crew can participate in prevention of foreign material contamination through participation in the GMPs controlling objects such as pens, jewelry, and buttons [8]. In addition, they can be observant of equipment condition, looking for loose or missing parts or damage from wear. Working in harmony with maintenance and production, sanitation can be a valuable asset in prevention of foreign inclusions.

References

1. Taylor, Dr. Steven L. and Dr. Sue L. Hefle, Allergen Control, *Food Technology*, 2005, pp. 40–43.
2. Cramer, Michael M., The Time Has Come for Clear Food Allergen Labeling, *Food Safety Magazine*, The Target Group, Glendale, 2004, pp. 18–22.
3. Deibel, Dr. Virginia and Laura Berkner Murphy, Writing and Implementing an Allergen Control Plan, *Food Safety Magazine*, The Target Group, Glendale, 2003/2004.
4. Graham, Donald J., Using Sanitary Design to Avoid HACCP Hazards and Allergen Contamination, *Food Safety Magazine*, The Target Group, Glendale, 2004.
5. U.S. Department of Agriculture, *FSIS Directive 8088.1*, Washington, D.C.
6. U.S. Department of Agriculture, *FSIS Notice 45-05, Verification of Activities Related to an Establishment's Controls for the Use of Ingredients of Public Health Concern*, Washington, 2005.
7. Wallen, P. and P. Haycock, *Foreign Body Prevention, Detection and Control*, Blackie Academic & Professional, London, 1998, p. 127.
8. Anon, Get the Lead Out, *Meat and Poultry*, 2005, pp. 53–56.
9. Joy, David, Uncertainty of Allergen Thresholds, *Food Processing*, August 2005, p. 21.
10. Hefle, Dr. Sue L., *Food Allergies and Intolerances*, presented at Food Allergens, Issues and Solutions for the Food Product Manufacturer, Chicago, July 29–30, 1998.
11. Gebhard, Dr. Roger L., *Gluten Sensitive Enteropathy*, presented at Food Allergens, Issues and Solutions for the Food Product Manufacturer, Chicago, July 29–30, 1998, pp. 1–5.
12. *Hahn, Martin J., Liability and Recall*, presented at Food Allergens, Issues and Solutions for the Food Product Manufacturer, Chicago, July 29–30, 1998.
13. Taylor, Dr. Steven L., *Food Allergens, Why Are They a Concern*, presented at Food Allergens, Issues and Solutions for the Food Product Manufacturer, Chicago, July 29–30, 1998.

Index